Planning for Sustainable Farming

Andrew Campbell, B.For.Sc. (Hons) was instrumental in developing and implementing the process of whole farm planning in his role as Project Manager of the Potter Farmland Plan from 1984 to 1988. He has since examined land degradation problems and the potential of farm and catchment planning all over Australia, firstly as Assistant Director of the Centre for Farm Planning and Land Management at the University of Melbourne and, since July 1989, as National Landcare Facilitator.

Andrew is still involved with his family farm at Cavendish in Western Victoria. His other interests include sport, photography, red gum furniture, bushwalking and cycle touring.

Planning for Sustainable Farming

The Potter Farmland Plan Story

Andrew Campbell

The colour plates in this publication have been made possible by
the support of Greening Australia, a community organisation dedicated
to helping all Australians to conserve and establish trees.

Lothian Books

A Lothian book

LOTHIAN PUBLISHING COMPANY PTY LTD
11 Munro Street, Port Melbourne, Victoria 3207

National Library of Australia
Cataloguing-in-publication data:

Campbell, Andrew, 1960–
    Planning for sustainable farming.

    ISBN 0 85091 433 7.

    1. Sustainable agriculture. 2. Sustainable agriculture –
    Case studies. I. Title.

630

Edited by Helen Chamberlin
Cover design by Lyn Twelftree
Text designed by R. T. J. Klinkhamer
Maps by John Ward
Illustrations by John Ward
Photographs by Andrew Campbell
Typeset in $^{11}/_{13}$ Berkeley O.S. by Midland Typesetters
Produced by Island Graphics
Printed in Australia by Impact Printing

Contents

Acknowledgements

This book refers mainly to the experience of developing the Potter Farmland Plan demonstration farms from 1984 to 1988. The farmers involved in the project and their families have all helped to generate this information and have taught me a great deal. I continue to be inspired by their efforts, and by many other good farmers around Australia, involved in farm planning and land conservation, whose interest and enthusiasm has spurred me to finish this book. Members of the Potter Farmland Plan Executive, particularly John Jack and Patricia Feilman, provided valuable advice, guidance and encouragement and made a significant contribution to early drafts of the text. Victoria Mack has been a great supporter of the project and contributed useful information on Hamilton 2000. Professor Carrick Chambers, Bill Middleton, Bill Sharp, Bob Piesse and Peter Dixon provided sound technical advice and valuable support. Finally, I owe an enormous debt to John and Sue Marriott, who put countless hours of meticulous effort into the project, from 1984 to 1988 and since. They have been a constant source of support, for which I am extremely grateful.

(1)

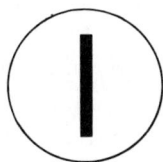

Introduction

MANY PEOPLE THINK of Australia as a young, lucky country blessed with an abundance of natural resources. As a western industrialised nation we may be young and developing, but our land is not. The Australian landscape is old and tired. 'Fragile' is a term which became hackneyed during the 1980s as a descriptor of our soils, water and vegetation.

It is not the soils, water and vegetation which are fragile, but our understanding of them.[1] The response of the Australian landscape to fire or climatic extremes exemplifies toughness and resilience, not fragility. It would have been impossible to convince the early settlers of the fragility of the Australian landscape as they battled against tenacious scrub, gutless soils and harsh climates, trying to farm as they had in Europe. Early accounts describe the unfamiliar land as vast, threatening and unknown, as needing to be 'tamed'. They portray the fledgling communities on the land as fragile, tottering between survival and bust. The creation of a large, diverse agricultural sector, in which 5 per cent of the workforce now generate 40 per cent of Australia's export income, is the result of the gut-busting hard work and resourceful ingenuity of generations of Australian farmers.

Through a combination of need, greed and ignorance, however, Australian agriculture has been living beyond its means. The inherent productivity of much of our farmland is very low. Before European settlement, stable communities of perennial plants and soft-footed animals survived by using very efficient methods of nutrient cycling and energy conservation. Vegetation provided a thin mantle of organic matter which protected the ancient, impoverished soils and nutrients were distributed from plants to the soil and back again in cycles which had been perfected over millions of years. The marsupials and other animals which so fascinated the European explorers have also evolved unique mechanisms for survival in Australian conditions, adapting to periodic dry spells which we now call 'drought' and making very efficient use of the subtle productivity of the Australian bush.

Our farming systems have often been inappropriate for Australian soils and for low, variable rainfall. Farming (and mining and forestry) has had an overwhelming impact on the animals and plants which had adapted over millions of years to these conditions. It is difficult to put into perspective the speed and violence of the impact of European man on the Australian landscape over the last 150 years, compared with the gradual changes which had imperceptibly moulded it for previous millennia. The axe, the plough, introduced crops and weeds, horses, dogs, cats, rabbits, sheep, cattle, goats, donkeys, camels, buffalo, insects, diseases and countless other immigrants have altered for ever the balances which had evolved between native animals and native vegetation. Lately the tractor, bulldozer, irrigator, boom spray and crop duster have greatly accelerated man's ability to mould the environment to cater for perceived immediate needs, whether to clear land for crops or pastures, to remove all competing pests from a chosen crop or to apply elements imported from other countries to bolster the fertility of hungry soils.

It is worth looking briefly at some of the stark impacts of Australian agriculture, mining and forestry on the landscape over the last 150 years:

- Two-thirds of Australia's forests (40 million hectares) and one-third of all scrub and woodland (63 million hectares) have been cleared.
- Forty-six mammal species (15 per cent of the total) have become extinct— a world-record rate of extinction.
- More than 500 species have been introduced as cultivars, weeds, pests or all three.
- More than half of all cropping and grazing lands require treatment for erosion, salting, soil acidity or soil structure decline.
- Many of the waterways and wetlands on this dry continent have become contaminated by soil run-off and algal blooms caused by fertilisers, pesticides and heavy metals.
- Most irrigation areas are being flooded from beneath by rising saline

groundwater, due to clearing, profligate water use and non-existent or inadequate drainage.

The annual cost in lost agricultural production of these problems was estimated by the CSIRO in 1989 dollars to be more than one billion dollars— we export $600 million worth of nitrogen, phosphorus and potassium each year in grain and livestock commodities alone. The immediate cost to arrest and repair this damage is estimated to be $2.5 billion, plus an annual cost of $180 million.

The extent of these impacts, and their long-term costs, have only recently been glimpsed by policy and decision makers in Australia and they are still only superficially recognised by the general community. Until recently, land degradation has not grabbed headlines. Soil is not cuddly or furry, nor rare and beautiful like reefs or rainforests. Soil conservation has taken longer to benefit from the growth in environmental awareness and concern which developed in the western world throughout the 1970s and 1980s. Finally, though, Australians are now worried about the basis of our liveli- hood—the land. Land degradation is, and is now recognised as, Australia's most serious environmental problem. In the early 1990s, efforts to correct past mistakes in land management are receiving unprecedented political and community support. It is encouraging that the groundswell of community concern for the environment and about land degradation in particular has not been accompanied by an atmosphere of blame. The entire Australian community, over seven or eight generations, has contributed to the land degradation problems illustrated by the few statistics above. Government regulations, prevailing community attitudes, the church, business, banks and the education system have all helped to develop the land-use patterns which have resulted in our current problems. At the same time, the entire community has built its present standard of living and lifestyle largely on the export earnings generated by the agricultural and mining sectors. All sectors of the community have contributed to these problems and it is in the community interest to fix them for the benefit of current and future generations.

This book suggests that it is not only possible to develop sustainable farming systems in Australia, but that we already have the necessary knowl- edge and examples from which to learn. It describes the development of a farm-planning process which can change the way in which farmers think about their land, using the practical experiences of case-study farmers from a demonstration farm project in western Victoria called the 'Potter Farmland Plan'.

But before we open the gate and enter the paddock of the farmer trying to support a family, in the face of rising costs and fluctuating returns, without degrading his land, it is important to discuss briefly this trendy notion of sustainability.

● SUSTAINABILITY

Sustainable development is becoming the catchphrase of the 1990s, both in agriculture and in the broader field of natural resource allocation and management. Sustainable agriculture is almost a hackneyed phrase, interpreted in a variety of ways according to the perspectives of its users. Broadly, sustainable farming systems must meet the needs of current generations without compromising the ability of future generations to meet *their* needs. Clearly, the degradation caused by the impact of modern agriculture on Australia's soils, water, plants and animals has already compromised current and future generations—most of our current farming systems are unsustainable by any definition. But simply to spend money to 'fix' the accumulated problems of the past 150 years with remedial measures is no more sustainable than to continue degrading the basic resource. The challenge now is to modify our land use and land management to prevent future degradation and, hopefully, to tackle existing problems at the same time.

Farming systems must also be economically viable, which means that the concept of sustainability embraces the balance between conservation and productivity—they are inextricable in the long term.

So what does sustainable agriculture really mean? Sustainable farming systems could also be called permanent or lasting farming, but to endure over time they must also be dynamic. If we are to learn from the natural systems which existed for the millennia before Europeans settled Australia, we must apply the essential elements of any natural ecosystem—control of populations, recycling of nutrients and efficient use of energy sources.

In the agricultural context, these can be combined to achieve the following definition:

> A sustainable farming system is one which is profitable and maintains the productive capacity of the land while minimising energy and resource use and optimising recycling of matter and nutrients.

Sustainability is obviously not an absolute state—it is a moving target for which we must continually adjust our aim with better knowledge and changing economic and physical circumstances. But it is possible to judge the relative degrees of sustainability of certain practices and systems, and to strive for improvement. An appropriate metaphor which helps to illuminate the concept of sustainability for me is that of a pilot flying a light aircraft. The pilot keeps the plane in the air and on target by observing a number of dials which indicate altitude, speed, fuel flow, trim, direction and so on, keeping all of these parameters within a safe range. There is no attempt to reduce all of these to a single gauge. Sustainable farming is similar, in that there are a number of indicators which the farmer needs to keep an eye on and to keep in balance, adjusting the farming system accordingly. These include soil structure and fertility, energy use, water and nutrient budgets, economic and ecological diversity and stability, and even social cohesion. As with the light aircraft, there is no single indicator of sustainability, but for each of the indicators used there is a direction which is more sustainable and we should be continually improving our farming systems in these directions.

For farmers who may have been farming in a certain way on a certain piece of land for many years this is a complex challenge. Sustainability is a vague notion and debate is often confused by interest groups using the term to further their own agendas. Modern communications and direct mail swamp farmers, as they do other sectors of the community. Assessing and absorbing this information and sorting out and applying what is valuable is very difficult for many farmers. It is often easier to carry on with 'business as usual'. This book concentrates on how farmers can learn more about their farming systems and their impact on the land and describes a process for planning change. The whole farm planning process explained in chapters 3 and 4 helps farmers to examine the management of each type of land on their farms and to implement improvements in an integrated way across the whole farm—to move towards sustainability.

● DEMONSTRATION FARMS—THE GERMINATION OF AN IDEA

There have always been farmers who have recognised the need to improve farming practices to prevent land degradation. Early steps towards more sustainable farming usually involved tree planting or reducing the amount of ploughing in preparing land for crops. Some farmers began to avoid the use of synthetic fertilisers and pesticides. These early pioneers were often regarded as eccentric or 'greeny' when it was not fashionable to be green. They learned many hard lessons which are now of value for the entire farming community. They were almost universally operating in a vacuum of supportive advice and information, swimming against the stream of information flow as well as prevailing community attitudes—and many still are. Naturally, some of the early tree planters and other conservation-minded farmers made mistakes, but valuable examples of what can be achieved were created.

The development of sustainable farming systems in Australia is further compounded by the way in which agricultural research and extension services have developed in Australia. A great deal of research has been done, research which has solved some of the problems of farming in such unfavourable conditions. Much of the research has achieved large increases in production and has contributed to our present high standard of living, but there has been a gradual specialisation of research effort as knowledge has become more detailed and specific. Extension services, responsible for assisting farmers to modify their attitudes and behaviour to apply new information and technology in improving farming systems, have also tended to become more specialised. Most departments of agriculture, for example, employ separate livestock or cropping officers, pastures specialists and economists, all dealing with their own aspect of the farming operation. In most states of Australia, soil conservation, wildlife and tree advisers are in different government departments and very few in number.

Australian farmers, on the other hand, are supreme generalists. They have a proven capacity to dramatically change their way of farming in a very short time, in comparison with farmers from other countries or with other sectors of the Australian economy. A farmer might be a mechanic in the morning, a veterinarian after lunch, a pasture manager in the afternoon and an accountant at night, planning the inputs for a diverse crop and livestock enterprise. But farmers are largely advised by specialists, and the information available to them is usually presented in bits and pieces within a particular discipline, rather than as a package which integrates information from various disciplines.

This reductionism has been accompanied by a separation of conservation and production in the minds of farmers and the rest of the community. With a few exceptions, the government agencies dealing with conserving the land resource are separate from those dealing with productive use of the land, which often leads to conflicting information and advice being disseminated by different departments. In 1977 the Hamer government recognised this when it created the Garden State Committee, under the Chairmanship of John Jack, then Assistant Secretary of the Department of Premier and Cabinet, a forester of 30 years' experience with a strong interest and background in ecology. The Garden State Committee was set up to achieve better coordination of government and private efforts in encouraging the conservation and establishment of native vegetation in Victoria and to develop an ethic of respect for the environment, and for its vegetation in particular, by promoting the image of the Garden State.

The Garden State Committee was made up of representatives of relevant government agencies, academia and private organisations. One of the first initiatives of the committee was to prepare a strategy to reverse tree decline in Victoria, which recognised the diverse range of contributing factors in rural tree decline and the need for an integrated approach to revegetation. It also recognised that local communities need to be directly involved in re-establishing vegetation, that it should not be seen as simply a matter for government to fix up, and that farmers can make best use of the limited advice available in groups rather than as individuals. After a highly successful 'Focus on Farm Trees' conference in Melbourne in 1980, the Garden State Committee, in conjunction with the then Victorian Farmers and Graziers Association, initiated farm-tree groups at Hamilton, Rochester, Wycheproof and Bairnsdale. Farm-tree groups involved farmers interested in conserving and establishing trees, with the addition of one or two representatives of relevant government departments, usually the then Forests Commission, Soil Conservation Authority or Department of Agriculture. It was intended from the outset that the groups would be led by and mainly comprise farmers, that they would be a forum for information gathering and dissemination about farm trees and that they would be active in promoting tree growing in rural areas. The farm-tree group programme grew quickly, and there are now fifty groups in Victoria, under the umbrella of the Victorian Farmers Federation and Greening Australia, which succeeded the Garden State Committee in 1989.

The farm-tree group programme illustrated the diverse sources of advice from which farmers must distil information relevant to their own situations. It also revealed a widespread belief within the farming community that it was not feasible to replace some of the vegetation which had been cleared— that trees need constant watering, that grubs eat them, that they just won't grow. In response to this, the Garden State Committee established demon-stration blocks of mainly indigenous vegetation at strategic sites around Victoria—usually on very exposed, degraded hills close to main roads, to show that even in the most difficult situations it is possible to re-establish trees. This very successful project was called 'Project Tree Cover'.

After showing that native trees could be successfully re-established (for a relatively modest cost) even in unforgiving conditions, a few members of the Garden State Committee were aware that many trees were still being planted in the wrong places, and often for the wrong reasons, in the Australian countryside. Indeed, trees were still being regarded as an issue separate from the rest of the farm enterprise. Professor Carrick Chambers was a member of the Garden State Committee and head of the Botany School at the University of Melbourne. He and John Jack, Chairman of the Garden State Committee, had often discussed these problems and each of them had launched successful farm-tree programmes in various parts of Victoria.

They mused on the benefits of setting up an entire farm to show how trees might fit into an ecologically sound farming system.

At about the same time, early in 1984, the Ian Potter Foundation, one of Australia's largest philanthropic trusts which was set up by Sir Ian Potter of stockbroking fame, was reviewing its programme. The Ian Potter Foundation had been a major contributor to the arts, conservation and medical research for over twenty years. There was now a feeling by some members that they should turn some of their attention to the problems affecting agriculture, particularly salinity. This was partly stimulated by an ABC television screening, over the weeks before this meeting, of the 'Heartlands' series of documentaries narrated by Dr Dean Graetz of the CSIRO. 'Heartlands' depicted in graphic detail the degradation and profound ecological imbalances occurring over Australian rangelands, awakening most of its viewers to these issues for the first time. Patricia Feilman, Executive Secretary of the Ian Potter Foundation, recalls that several of the Governors of the Foundation had seen the programme and were moved by it to suggest that the Foundation should sponsor an effort to tackle land degradation. She was asked to investigate how the Foundation could best contribute.

Pat Feilman is a trained accountant who has an extensive and varied background in the business world and in public interests. When she was asked to develop a project relating to land degradation, Pat was President of the Nurserymen's Association of Victoria and was also a member of the Zoological Board of Victoria, which she later chaired. She also became the second Chairman of the Garden State Committee, succeeding John Jack in 1987. Through these and other interests, Pat had known Carrick Chambers for some years, and she asked him for advice as to how the Foundation might get involved. He recommended that the time was right for establishing demonstration farms that could illustrate to agriculturalists that the reversal of land degradation could be tackled at its root cause by blending agriculture with applied ecology. Clearly there was a powerful group of people wanting to help. The Ian Potter Foundation and the Garden State Committee (with the support of the Victorian Government) agreed to work together to develop a project. The Potter Farmland Plan was born.

When Carrick Chambers and John Jack began to discuss the establishment of some demonstration farms, they focused immediately on the areas where the first farm-tree groups emerged, reasoning that these areas have farmers with a proven interest in tackling land degradation and, at the same time, returning trees to their farm landscapes. They recommended the Dundas Tablelands, near which the Glenelg Farm Tree Group was established at Hamilton, as an ideal demonstration region.

The Dundas Tablelands are a prominent feature of the country explored

• 'Helm View' is an outstanding example of a property which has substantially implemented a whole farm plan. Water supply, fire protection, farm layout, pasture composition and management, landscape values, wildlife habitat, shelter for stock and crops, drainage and land degradation control have all been integrated in the plan and in the paddock, resulting in a more stable, attractive and productive environment in which to live and work.

• An attractive vista from just south of Glenthompson. Land degradation is often a subtle process, slow in terms of normal human planning horizons. Changes need to be made to the farming systems in this area to achieve a better water balance, to improve soil stability and fertility (and profitability) and to preserve biological diversity, if views like this are to be possible for future generations.

• Before and after: the top photograph shows an old fence dissecting a waterlogged, salt-affected drainage line. The lower photograph, taken five years later, shows the low-lying area fenced into one unit, sown to salt-tolerant perennial grasses and surrounded by direct-seeded trees and shrubs. The production from this area has more than quadrupled and it is now a farm asset.

by Major Thomas Mitchell in 1836, which he called 'Australia Felix', describing it thus: 'A finer country could scarcely be imagined'.

These words and other descriptions in Mitchell's report caused a rapid influx of settlers, whose impact on the land will be described later. First, a bit more about its original condition.

On 12 July 1836, Major Mitchell climbed the highest peak of the range he named the Grampians and decided to continue his expedition westwards, rather than south along the eastern edge of the mountains, as the ground appeared too soft. By 31 July the party had travelled west, hoping to follow the Wimmera River to its mouth, which they assumed to be along the south-east coast of what is now South Australia. Climbing Mt Arapiles on 23 July, Mitchell observed that the Wimmera turned to the north-west, 'leaving me in a country covered with circular lakes, in all of which the water was salt or brackish'.

Mitchell and his assistant, Granville Stapylton, concluded that, 'it is reasonable to suppose that the salt water on these Lakes has been left there by the receding of the Ocean'.[2] They were right. The expedition turned south, crossing the Glenelg near the present town of Harrow, and following it down as far as Dartmoor before veering south-east to Portland by 30 August where, to their surprise and disappointment, the Henty family had already established a farming and whaling settlement. As they crossed the tributaries of the Glenelg on their journey south, they were see-sawing over the alternate flat-topped hills and steeply dissected valleys which we now call the Dundas Tablelands. An extract from Stapylton's journal of 29 July 1836 captures their enthusiasm for this landscape (I have inserted punctuation, absent from the original):

> A running Creek, the first we have seen for many days, course South East, the very thing we wished, High Forest ranges bordering it right and left. The descent of the Creek is remarkably sudden, 1 mile from its source it becomes A Large Tributary to some great River ahead. Its rapid increases in bulk and the sudden descent from the Low level Lake Country implies our proximity to the Ocean. Primitive Rock on the ranges which are high and dry. Foliage of the Timber rich and shady, in short the most picturesque Forest Land I have ever seen. Noble and wide ranges, grassy and thinly covered with Swamp Oak, Banksia and Lofty Gums. Such A situation would be invaluable as A grant coupled with the illimitable run for Cattle about the Salt Lakes, where also innumerable fresh water Swamps abound. Nothing could have happened more unexpectedly than this sudden change from Deep Flats to good granite ranges . . .

Three days later Stapylton described the area which they named Pigeon Ponds: 'the forest ranges richly clothed with grass xanthonia in the valleys positively having the appearance of a crop of oats'.

Stapylton's 'xanthonia' is probably the perennial tussock grass Danthonia, which was accompanied by Kangaroo Grass (*Themeda australis*), Spear Grass (*Stipa* species) and *Poa* species. The forests admired by Mitchell and Stapylton were woodland and open forests comprising mainly River Red Gums (*Eucalyptus camaldulensis*) along the drainage lines and flats, Manna Gums (*Eucalyptus viminalis*) and Swamp Gums (*Eucalyptus ovata*) on the slopes and tablelands, with an understorey of smaller trees such as Silver Banksia (*Banksia marginata*), Blackwood (*Acacia melanoxylon*), Black Wattle (*Acacia mearnsii*), Drooping Sheoak (*Allocasuarina verticillata*) and Sweet Bursaria (*Bursaria spinosa*). Further down the Glenelg, Mitchell describes Stringy-bark (probably *Eucalyptus obliqua*) and Blue Gum (*Eucalyptus globulus*) 'some as much as eight feet in diameter'.

Stapylton's term, 'an illimitable run for cattle', seems to have been taken literally by the energetic Hentys, as they were the first of many to spread their stock into Australia Felix after hearing tales of promise from Mitchell's party. In August 1837 the first mob of sheep was driven to the Merino Downs and by winter of 1840 land as far as Chetwynd had been taken up by squatters including Winter, Bryan and John G. Robertson.[3]

The first clearing of the Dundas Tablelands and areas south and east was probably limited to the immediate vicinity of homesteads (such as they were) and tracks. But the impact of sheep was immediate, profound and widespread, as they took with an indiscriminate vengeance to the lovely perennial tussocks and their multitude of cloven hooves ground the reluctant soil into submission. As early as 1853, the John Robertson mentioned above was moved to write to Governor La Trobe about conditions at Wando Vale, in an often-quoted extract:

> When I arrived through the thick forest land from Portland to the edge of the Wannon country, I cannot express the joy I felt at seeing such a splendid country before me... The few sheep made little impression on the face of the country for three or four years... Many of our herbaceous plants began to disappear from the pasture land... the ground is now exposed to the sun... the clay hills are slipping in all directions... when I first came I knew but two landslips, now there are hundreds found within the last three years... springs of salt water are bursting out in every hollow or watercourse. Strong tussocky grasses die before it, with all the others... when rain falls it runs off the hard ground... into the larger creeks and is carrying earth, trees and all before it... Over Wannon country it is now as difficult to ride as if it were fenced. Runs seven, eight and ten feet deep are found for miles where two years ago it was covered with tussocky grass like a land marsh.[4]

Robertson's communication to La Trobe also illustrates that another characteristic of Australian agriculture—the cost/price squeeze, also became evident very early after settlement:

> About twenty of the squatters in the Portland Bay district were sold off. Three or four I knew compromised for less than half with their creditors—there is not one station I know but my own and two others that is occupied by the original squatter.[4]

Sheep were worth £2 to £3 per head in 1828, but in the depression of the 1840s, were worth less than five shillings. Robertson mentions that in 1843 he purchased (for his cousin) Warrock Station on the Glenelg, including a team of bullocks, 2500 sheep and improvements, for a total cost of £300. It had cost its founders £5700 to establish and keep for three years. The people who survived such tough times, often with only rough slab buildings for shelter, were people of capital, fortitude or both. It is estimated that nearly two-thirds of the pioneer settlers of the region were from the Scottish lowlands and had a farming tradition of centuries behind them.

The prosperous, stable farming community of western Victoria today owes a great deal to the resourceful stoicism of these pioneers, but we also owe a debt to the land. By 1903 Bruni remarked of the land around Hamilton that:

> ... since the squatting days fully three-quarters of the timber has disappeared ... the country is becoming so open that ere long the landholders will have to set about establishing shelter plantations.

One hundred and fifty years after Major Mitchell the problems caused by overclearing, overstocking and rabbits were recognised to the extent that the Victorian government established a parliamentary inquiry into salinity. The Hamilton public hearing of the inquiry in July 1983 was told that up to 5 per cent of the Dundas Tablelands was affected by dry-land salinity and that gully, sheet and tunnel erosion had continued to get worse since Robertson's first report in 1853. These problems were accompanied by a disturbing decline in the health and appearance of many of the remaining paddock trees, the characteristic grandeur of the 'Red-Gum country' which was so much admired, particularly in evening light.

Some far-sighted farmers had recognised these problems and were doing something about them long before the politicians. Neil Lawrance, farming at Gatum, north of Cavendish, fenced out his first salt patch and rehabilitated it by mulching it with old hay and sowing it to salt-tolerant grasses and trees in 1960. John Fenton at Branxholme committed himself to a long-term programme of tree planting and returning water to natural wetlands for wildlife habitat and water conservation early in the 1960s. Over the next decade, several other local landholders—notably Richard Jamieson from Woorndoo, Richard Weatherly from Mortlake, Robin Jackson from Lake Repose and the Macgugan family from Grassdale—made significant advances with farm tree establishment. When the farm-tree group programme began, Glenelg was an obvious first.

Thus, returning to the point of the story, when John Jack and Carrick Chambers and others put forward the Dundas Tablelands as a site for a demonstration project, they were considering a region which could display all the main types of land degradation in abundance within a reasonable radius of Hamilton, but which also contained some environmentally aware farmers who had already shown that they were prepared to confront land degradation.

In early to mid 1984, staff from the Garden State Committee (mainly Executive Officer Alan Thatcher and Project Officer Stephen Farrell) prepared a detailed outline of the proposed project under the supervision of John Jack, with inputs from Carrick Chambers, Pat Feilman and Bill Middleton. Bill Middleton is a forester and skilled naturalist with 30 years' experience (mostly in the Wimmera and Western Victoria), running the Forests Commission's Wail Nursery and promoting the establishment of trees on farms and the protection of ecologically critical remnant vegetation. For sixteen years, in the 1960s and 70s, Bill was 'the Western Victorian Gardener' on ABC radio, hosting weekly chats about the ecological side of farming, long before it was fashionable to be interested in native plants and birds on farms.

On 7 and 8 August 1984 representatives of the Garden State Committee, the Ian Potter Foundation, the Glenelg Farm Tree Group, the then Victorian Farmers and Graziers Association, the Department of Agriculture, the then Soil Conservation Authority and the then Forests Commission made an extensive tour around the Dundas Tablelands region. They identified three distinctly different localities, each with its own farm types and land degradation problems, which would be suitable for demonstration purposes. They were Wando Vale, Melville Forest and Glenthompson (see page 22).

● THE SETTING

The Wando Vale demonstration area (80 kilometres west of Hamilton) is on the steeply dissected western side of the Dundas Tablelands. It is beautiful country, still deserving of the glowing descriptions it inspired in Mitchell and Stapylton, with steep slopes, delightful permanent streams and flat, wooded hilltops forming a landscape which is much more visually interesting than most Australian farmland. The sandy loam topsoils often have outcrops of 'buckshot' or ironstone gravel concretions on the ridges, and the subsoils are yellow clays on the slopes and black clays on the valley floors. As John Robertson noted in 1853, the area is highly prone to gullying, tunnel erosion, saline soaks and landslips (where soil slips down a hill after the subsoil becomes saturated and unable to support it). Tree decline is becoming more severe among the Red Gum (*Eucalyptus camaldulensis*), Manna Gum (*Eucalyptus viminalis*) and Snow Gum (*Eucalyptus pauciflora*) remnants.

Most farms in the Wando Vale area are grazing properties of 400-800 hectares, running sheep and cattle, with a few smaller dairy farms in the valleys close to Casterton and a few larger sheep and cattle properties of more than 2000 hectares. The climate varies even within one farm, from the tableland country, which is over 400 metres above sea level and has occasional severe winds and frosts, down to the valleys which are milder. The average rainfall varies from 700 to 800 millimetres per year, and is very reliable, making this a prime farming area. Lucerne is grown for high-quality hay by some farmers on favourable banks and occasional crops are grown, mainly for stock feed.

Melville Forest (45 kilometres north of Hamilton) is representative of the beautiful 'Red Gum country' of the eastern Dundas Tablelands, on soils derived from rhyolite. The average annual rainfall for Melville Forest is a reliable 650-675 millimetres, and the farms are mainly sheep enterprises of 200-600 hectares, with few farms larger than 1000 hectares. The classic Red Gum (*Eucalyptus camaldulensis*) savannah woodland has been heavily modified since settlement. Most of the original understorey of Blackwood and Sheoak has been cleared and is now confined to roadsides. The area is subject to increasing dry-land salinity and erosion.

Melville Forest was settled in the 1840s by sheep farmers on extensive runs, which gradually became smaller over the next fifty years of closer settlement. In a bizarre move typical of many bureaucratic decisions affecting land use in Australian history, Melville Forest was the site of an attempt to settle English soldiers on farms after the Boer war. Bruce Milne's father recalls that these chaps led the 'life of Riley' for a few years—hunting, dressing for dinner, playing croquet and other gentlemanly pursuits. But they were not farmers, and no one had taught them how to farm. One by one they walked off with nothing.

It is hard to see why Melville Forest is so named, as the few grand but tired old Red Gums dotting the landscape resemble anything but a forest. When Peter Waldron's grandfather came to the area at the turn of the century, the crowns of the trees were touching. My grandmother tells the story of the first white settlers to the district 10 kilometres north of Cavendish, where my family farm is today, who needed axemen to clear a path for the dray through the forest. We now have less than one large Red Gum per acre, and they are dying at an alarming rate, due to 'AIDS'—Age, Insects, Disease and Stress. At one stage Melville Forest boasted numerous sawmills, cutting Red Gum for railway sleepers, bridge decking and the blocks which paved the streets and tramways of Melbourne and Adelaide; many of these are still evident today.

The timber cutting took place from the turn of the century through to the 1930s. Tall, straight trees were felled and milled, and most of the crooked ones were ring-barked to kill them. Many of these skeletons are still standing, with the necklace of axe cuts around the trunk and stark grey limbs stretching skyward, prompting one visitor to describe them as 'strangled screams'. The early settlers thought that the trees competed with the pasture, that they caused liver fluke in sheep, and that they made rabbit control more difficult. The last point was true then, but it is no longer valid today. During the depression of the early 1930s Peter Waldron's father was employed ring-barking for sixpence per tree, which was good money in those hard times.

We can smile at or wistfully regret the mistakes of the old-timers, but they were operating according to the conventional wisdom and prevailing community attitudes of the day—not just among farmers, but throughout the community. Clearing was not only good, it was an essential pastime for good farmers. The communal image of the 'good farm' was one laid out on the square like a chequerboard, cleared from fence to fence, with perhaps some conifer hedgerows to accentuate the regularity, and some deciduous trees in a formal avenue along the driveway for colour in the autumn. These attitudes die hard, and are still prevalent today, encouraged and exemplified by legislation. It was only during the 1980s that the tax concessions for clearing were removed!

I prefer not to dwell on the mistakes the old-timers made, but to applaud what they were able to achieve with the knowledge and equipment available. Anyone who has ever tried to remove just one overgrown eucalypt from a backyard should admire a generation who cleared so much tenacious scrub with axes, bullocks and chains. They did a mighty job and achieved what they set out to do, because they had the full weight of public opinion and government (such as it was) behind them and they persevered. There are some valuable lessons there for us as we try to repair some of the damage and to develop farming systems which do not degrade the land. We have the benefit of hindsight and new techniques which the pioneers did not have. More of that later.

As in other farming districts, it wasn't just the big trees which were cut down or killed at Melville Forest. Most of the original understorey of Blackwood (*Acacia melanoxylon*) and Sheoak (*Allocasuarina verticillata*), together with a range of associated shrubby and herbaceous species, was cleared and what is left is now confined to roadsides. The area seems to have some of the most severe dieback of the remnant Red Gums in western Victoria. Most creeks in the district are salty, and some dams are now completely useless for stock because of salinity. Most farms have land along their drainage lines which is now too saline for pasture grasses to grow, and these saline areas are commonly around 5 per cent of the total farm area. In many of these salty creeks soil structure has deteriorated, allowing the banks to be eroded by water and the hooves of sheep and cattle. The thirteen sawmills, the ring-barkers and generations of sheep, cattle and rabbits since have taken their toll. The countryside is still beautiful, but to an informed eye, it is sick.

The Glenthompson area (50 kilometres east of Hamilton) is typical of much of the western district today, with rolling, windswept country broken only by Sugar Gum (*Eucalyptus cladocalyx*) or Cypress (*Cupressus macrocarpa* or *C. lambertiana*) shelter belts. The area is high in the western reaches of the Hopkins River, which runs from the Pyrenees in the north to Bass Strait at Warrnambool and is now suffering from salinity problems. The geology of the area around the demonstration farms is varied, as deeply weathered Devonian granodiorites have intruded into Ordovician shales and sand-stones.[5] Crops and pastures are grown on 'texture contrast' soils (where there is a sharp boundary and contrast between the texture of the topsoil and the subsoil), usually sand or sandy grey loam topsoils over yellow clay subsoils. The average annual rainfall is 600-700 mm. Farms in the demon-stration area are subject to very severe gully erosion, wind erosion on the higher sandy slopes, tree decline among the very few remnants and increasing salinity. Some farms have more than 10 per cent of their land affected by salinity, and the gullies are often more than 5 metres deep and 5 metres wide.

The Glenthompson area was settled at about the same time as the Wando Vale and Melville Forest districts, and has so far followed a similar pattern, although farms are larger—averaging from 400–1000 hectares, with a few over 2000 hectares. Sheep and cattle are the main enterprises, although cropping (usually oats and lupins) is more common than in the other two areas.

---

[1] Ted Lefroy of Western Australia explains this beautifully: 'Soil, water and vegetation are anything but fragile—they are as robust and adaptable as the processes occurring within them as long as we respect those processes. If we know the physical limits of an environment (as described by climatic and edaphic parameters) and we recognise the range of processes that can occur within those limits, then we should be amazed at how tough any landscape is. As an example I recently watched a two year old destroy a

state-of-the-art video recorder in thirty seconds. The video recorder wasn't fragile, it just wasn't designed to have a three-inch nail and a piece of toast posted through its cassette hole. Similarly, a bulldozer is not fragile because it requires a complete overhaul when someone puts a few grams of compound X in its fuel tank.'

2 Alan E. J. Andrews (ed.), *Stapylton, With Major Mitchell's Australia Felix Expedition, 1836*, Blubber Press, Hobart, 1986.

3 Andrews, p. 171.

4 J. F. Bride, *Letters from the Victorian Pioneers*, Trustees of the Public Library of Victoria, Melbourne, 1898.

5 P. Dixon, 'Dryland Salinity in a Subcatchment at Glenthompson, Victoria', *Australian Geographer* 20 (2), November 1989.

2

Getting Organised

● ADMINISTRATIVE AND ADVISORY STRUCTURE

B Y SEPTEMBER 1984, planning of the as yet unnamed project was well under way in Melbourne and its broad focus and areas of operation were defined. An executive committee was established comprising:

Chairman	John Jack
Ian Potter Foundation	Patricia Feilman
Land Protection Division, CFL	Bob Campbell, Dennis Cahill, 1987–88
Victorian Farmers Federation	Doug Richens, Bob Carraill, 1986–88
	John Diprose
University of Melbourne	Ian Ferguson
Victoria Conservation Trust	Ian Wilton, Warwick Forge, 1986–
	Bill Middleton
Communications consultant	Peter Mathews
Department of Agriculture	Jim McLaughlin, Bruce Muir, 1988–
Senior Field Advisor	Bill Middleton
Public relations consultant	Mr Derek Sawer

The executive, based in Melbourne, was supported by a project officer and, from early 1985 to mid-1989, a part-time administrative officer.

I was appointed Project Manager in October 1984. I grew up on a farm north of Cavendish, less than 20 kilometres from the Melville Forest area as the crow flies, and my family has been farming in the area for generations. I studied Forestry at Creswick and the University of Melbourne and worked with the Forests Commission in farm-tree extension, forest management and fire suppression in the Grampians and at Bendigo and Shepparton. It became obvious very early in my forestry career that the most exciting and rewarding area in which I could apply my training was in working with farmers in revegetation. After a great deal of contact with farmers through the Forests Commission's Tree Growing Assistance Scheme I was convinced that for best long-term results, it is essential that trees on farms are planned as an integral part of the whole farm operation, not 'bunged in' as an afterthought.

The position of Project Manager for a demonstration-farm project in Western Victoria, to be funded by the Ian Potter Foundation, was advertised within the Department of Conservation, Forests and Lands (CFL) in spring 1984. This department was formed in mid–1984 from the amalgamation of the Forests Commission, the Soil Conservation Authority, the National Parks Service, the Lands Department, and Fisheries and Wildlife. It became the Department of Conservation and Environment (DCE) in 1990.

I thought such a job sounded too good to be true. A detailed Project Manager's brief had been prepared by Alan Thatcher and Steven Farrell of the Garden State Committee for the project executive, setting out broad project aims, proposed activities and anticipated outcomes at the end of three months, six months, one year and three years. Throughout the project, the management structure and decision-making processes were streamlined and effective. As Project Manager, I was responsible directly to the Melbourne-based Executive. The Project Manager was required to present quarterly reports to the Executive, documenting the progress of operations at Hamilton, summarising expenditure, making recommendations and seeking approval for major items of expenditure and any changes of direction or new initiatives. The contrast between our operation and the government office in which we were located was stark. I could write cheques for equipment and other local purchases and we could reliably negotiate with local suppliers knowing that we could pay within seven days, which is a real asset in a country town. We had a general policy of purchasing everything locally if possible, even to the extent of getting some publications printed in Hamilton rather than Melbourne.

It was envisaged from the start that a Works Supervisor would be appointed to assist the Project Manager, particularly with on-ground works and documentation of works activities. John Marriott was lured from his dairy farm at Myamyn (near Heywood in south-west Victoria) to this position in January 1985. John has a very strong farming background, having managed large grazing properties in western Victoria for many years prior to developing his own farm. He has a Diploma of Agriculture from Lincoln College in New Zealand and by 1985 he had developed excellent skills in drafting and had a sound technical knowledge and practical experience in revegetation after years of part-time work for the Forests Commission in Heywood. In recognition of the evolution of his role from works supervision to extension as the project developed, John's title later changed to Project Officer.

During the first six months of the project, from November 1984 to May 1985, CFL provided technical assistance to the project through the full-time involvement of Bill Middleton, in the position of Senior Field Advisor, and the part-time involvement of Bill Sharp, Soil Conservation Advisor. On a local level, the Project Manager and the Project Officer were advised by a Local Advisory Group, chaired by Bill Middleton, comprising:

Victorian Farmers Federation (VFF)	Stuart Cuming
Department of Agriculture and Rural Affairs (DARA)	John Heath, Banjo Patterson, 1986, Bruce Knee, 1987–
CFL	Bill Sharp, John Langford 1985–
Glenelg Farm Trees Group	Richard Jamieson
Participating farmers from each demonstration area	

Originally three farmers, one representing each area, were on the Local Advisory Group, but in August 1987 the membership was extended to include all demonstration farmers. Demonstration farmers included the principal partner in each operation, and those directly involved on the demonstration farm, including wives and children.

The Local Advisory Group advised the Project Manager on aspects of demonstration work, assistance to associate demonstration farmers, and local public relations and extension, such as field days. It also provided an opportunity for the demonstration farmers to express any concerns and was the main forum for group interaction among the demonstration farms.

● SELECTING FARMS AND FARMERS

In December 1984 Peter Mathews, in his role as communications consultant to the project executive, organised a 'consultation' at Creswick to introduce the germinating idea of the project to farmers from the demonstration region and representatives of other sectors of the community and to give participants a chance to listen to all points of view, to contribute and to take responsibility for decisions made. Peter had much experience in group consultation from his career with Australian Frontier, and his facilitating skills with groups of people from a diverse range of backgrounds made a tremendous contribution as the project evolved.

The consultation process provided a neutral meeting ground, an independent chairman, an open agenda, and no press reports or formal resolutions. The Creswick consultation gave about twenty farmers from the Hamilton region the opportunity to meet local bankers, accountants, government officers, academics, educators and representatives of the project executive, to hear the Potter proposal and consider if they felt it worth putting to other farmers in their district. The farmers had been nominated by the Hamilton office of the Department of Agriculture, the Soil Conservation Authority, the Glenelg Farm Trees Group and the then Victorian Farmers and Graziers Association as 'opinion leaders' within the farming community.

The Creswick consultation was a watershed in the development of the project. The participating farmers were enthusiastic, the range of people present began to see its tremendous potential and to feel a responsibility towards the project by having been involved in planning at such an early stage in its development. Participants agreed that the project had merit, and went through an interesting exercise of naming the embryo—eventually coming up with 'Potter Farmland Plan'. Farmers were particularly keen to see the contribution of the Ian Potter Foundation reflected in the name, the emphasis on planning had to be conveyed, and the sense of a broader ecological approach was embodied in the use of the word 'farmland', rather than just 'farm'.

The Creswick consultation also approved the criteria for the selection of farms as follows:

FARM SELECTION CRITERIA

- Farms must fall within the three delineated areas.
- Farms should preferably be located high in a catchment, so that works carried out would be as independent as possible from any flow-on effects of poor management on other properties within the catchment.
- Farms should preferably have a subcatchment within the property.
- Farms should have significant evidence of land degradation.
- Farms should be readily accessible and preferably in highly visible positions.
- Farms should have been managed on sound agricultural principles, so

that the project was seen to address issues common to the region rather than the effects of a history of poor management.
- The range of farms selected should reflect the diversity of enterprises, land types and management approaches in the region.
- There should be evidence that the farmer has made a real effort to address problems.

REQUIREMENTS OF PARTICIPATING FARMERS

- Contribution in cash, equipment, materials and labour of at least one-third of the total cost of implementation and maintenance of the whole-farm plan. This contribution could vary on an annual basis as agreed between the Executive and the farmer.
- Willingness to allow access to and use of the property for monitoring, demonstration and publicity purposes indefinitely.
- Completion and maintenance of the works as outlined in the whole-farm plan.
- Willingness to keep a financial and physical record as required by the Executive.
- Willingness to cooperate with the Potter Farmland Plan Executive in the development of the project.
- Signature on a formal agreement covering the conditions above.
- Subsequent agreement that, should the farm be sold, the existence of the 'plan' and its fundamentals would be brought to the notice of pro-spective purchasers.

The Creswick Consultation also agreed that all landholders whose prin-cipal dwelling was within each demonstration area would be invited in writing to a public meeting. Demonstration farms were to be selected from volunteers after the meetings. The selection criteria, and an introduction to the aims of the Potter Farmland Plan, were to be circulated with the in-vitation to the meeting. A sub-committee comprising John Jack, Bill Middleton and Andrew Campbell was empowered to select a minimum of two and a maximum of six farms from each of the three demonstration areas.

The three public meetings were held in January 1985. About eighty farmers attended and forty-five applied to be involved in the project. All the applicants were interviewed, their farms were inspected by the committee in mid January and a short list of potential demonstration farms was prepared according to the criteria established at Creswick. Those farms on the short list were inspected again, and discussions were held with landholders about their

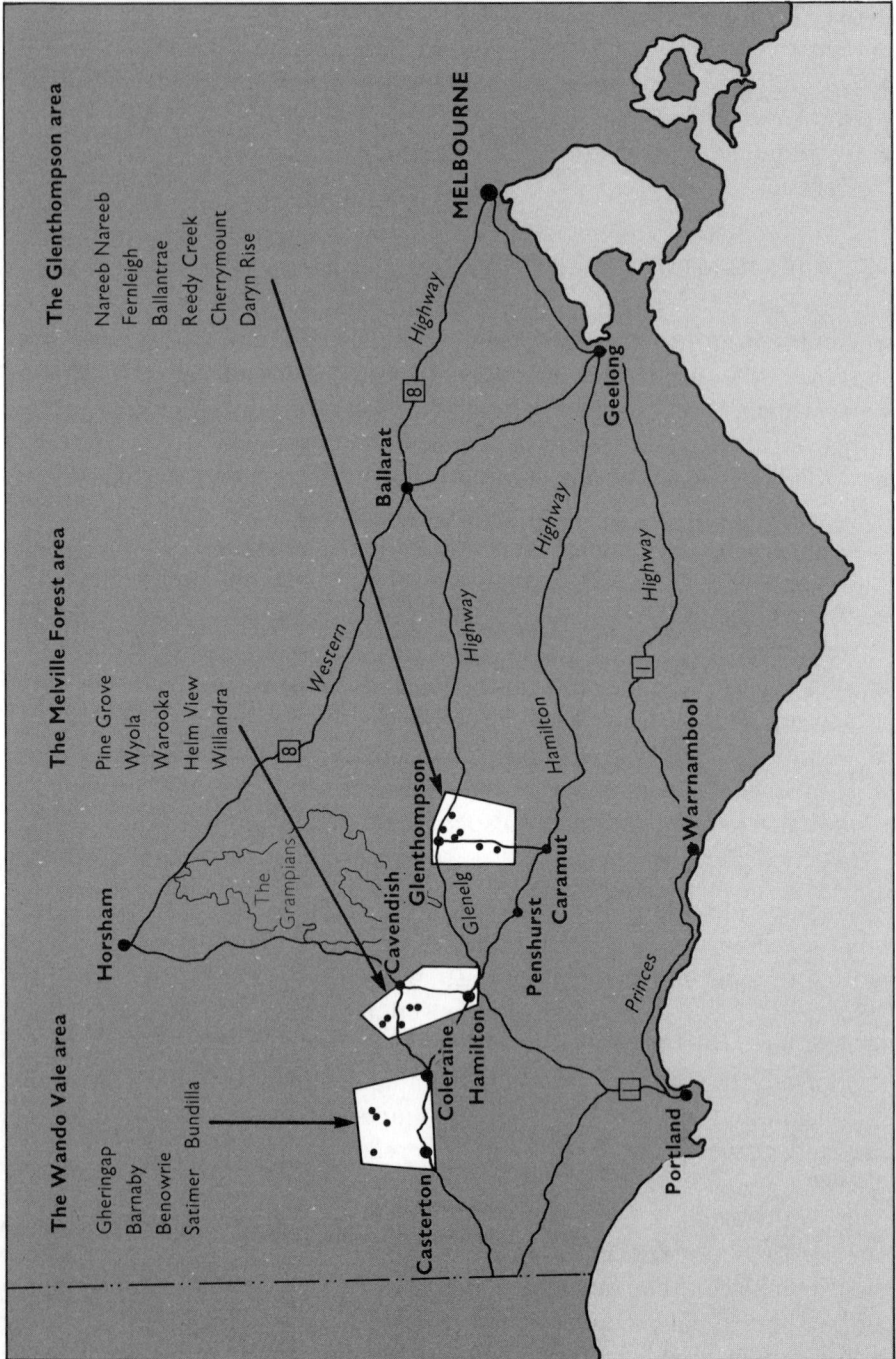

The Wando Vale area

Gheringap
Barnaby
Benowrie
Satimer Bundilla

The Melville Forest area

Pine Grove
Wyola
Warooka
Helm View
Willandra

The Glenthompson area

Nareeb Nareeb
Fernleigh
Ballantrae
Reedy Creek
Cherrymount
Daryn Rise

MELBOURNE
Geelong
Ballarat
Horsham
Glenthompson
Cavendish
Penshurst
Caramut
Warrnambool
Coleraine
Hamilton
Casterton
Portland

The Grampians

Western Highway
Hamilton Highway
Highway
Princes Highway
Glenelg

• The location of the Potter Farmland Plan demonstration farms.

potential contributions to the project. Sixteen demonstration farms (six at Glenthompson, five at Melville Forest and five at Wando Vale) were selected and recommended to the Executive for approval in February 1985. One of the landholders selected at Wando Vale withdrew his application and fifteen farms were finally approved by the Executive in April 1985.

● THE FARMS

While the demonstration farms as a group were supposed to represent the range of farm sizes, farm enterprises (the type of business run on the farm— for example wool growing, cropping, dairying) and land types, at no stage was it intended that the farms would be hand-picked to achieve a rep- resentative sample. The Potter Farmland Plan demonstration farmers were selected from a group of people who *volunteered* their farms, and a substantial amount of their time and money, without really knowing what the project would mean for themselves or their land. There was no 'recipe' or established norm for what we were proposing to do. Farmers were to be assisted to develop and implement their own farm plans, which were to integrate the agricultural demands of the farm business with the ecological needs of the land, but apart from these broad parameters the project was to evolve as it went along.

The farmers who applied to be involved tended to be younger, better educated and with larger farm sizes and a longer length of ownership than the average for the district—typical 'risk takers' in the jargon of extension. The farms chosen range in size from less than 200 hectares to over 3000 hectares and cover the full range of land types over the three demonstration areas.

The 'before' picture of each farm is briefly sketched below, but a fuller picture will emerge as the impact of the project is described over the coming chapters.

WANDO VALE

'Gheringap'

Ross and Annabelle Kitchin have owned 'Gheringap' since 1981. The pro- perty was an unimproved block of 260 hectares when the Kitchins took over, poorly subdivided into large paddocks, with poor annual pastures and almost no fertiliser history. It is a very attractive block on Satimer Road, approximately 25 kilometres north-east of Casterton, with healthy stands of indigenous eucalyptus, acacia and casuarina species (albeit with a lack of understorey species due to grazing), particularly on the steeper slopes. It is dissected by the Corea Creek (a tributary of the Wando River), and is typical of the area, with some flat tableland broken by steep escarpments and very clearly defined, often deep, gullies. It has several active soaks, and some very active gully erosion caused by stock traffic to the creeks, which were the main water supply prior to 1985.

Ross and Annabelle run a self-replacing Merino sheep flock, and had embarked on a programme of farm development involving new pastures, new fences and better subdivision (including laneways) when they applied to be involved in the Potter Farmland Plan. During 1987, they purchased 300 hectares of the neighbour's property to the north, creating a compact block of 560 hectares, which is a viable property size for the area.

'Barnaby'

Jeremy and Jill Lewis moved to 'Barnaby', on Satimer Road in 1978. They operate a Perendale sheep flock in partnership with William and Jennifer McKellar, who live in Geelong. The 650-hectare property is dissected by the Wando River, which runs from north-east to south-west, and it is notable for two 10-hectare areas of remnant woodland comprising Manna Gum (*Eucalyptus viminalis*), Snow Gum (*Eucalyptus pauciflora*) and Drooping Sheoak (*Allocasuarina verticillata*), with local understorey species such as Blackwood (*Acacia melanoxylon*), Black Wattle (*A. mearnsii*), Native Cherry (*Exocarpus cupressiformis*) and Sweet Bursaria (*Bursaria spinosa*). The northern two-thirds of the farm is a flat tableland which drops steeply to the Wando and the southern portion slopes more gently from Satimer Road to the river.

When purchased by the Lewis partnership, 'Barnaby' was relatively un-improved, with very large paddocks, poor pasture and poor fencing. The steep slopes along the Wando were subject to landslips and parts of the river bank were affected by gully erosion and rabbits. Some of these areas had been fenced out by the Soil Conservation Authority in the past. Stock camps on high ground were also a problem, typical of many throughout the district.

'Benowrie'

Peter Sandow owned 'Benowrie', a 972–hectare farm on the Casterton–Eden-hope Road, 20 kilometres north of Casterton, from 1964 to early 1988. 'Ben-owrie' is largely undulating, with several steep gullies, through which Deep Creek and Vines Creek run, eventually finding their way into the Glenelg

• Peter Waldron has converted the dead, dry old Red Gum trees and posts on 'Willandra' into attractive rustic furniture. This lucrative sideline gives him a buffer against falling commodity prices and provides him with the cash to take land out of short-term production for the next crop of trees. Old fence posts are no longer rubbish or firewood!

• Before and after: this direct-seeded octagonal clump of Red Gums on 'Willandra' now obscures the salt-affected dam behind it, which has also been fenced off and revegetated. It will be desirable in the future to thin out this clump for posts and poles, and also to seed some shrubs into it as an understorey.

• The partly low-lying pasture above was useless Sea Barley Grass in 1985. By 1988 it had been sown to more productive salt-tolerant species. The low-lying area was fenced separately to allow more sensitive management and trees were established to turn these small paddocks into well-sheltered areas. The lateral spread of the salt seems to have stopped.

• Before and after: it is almost too late to appreciate the full impact of the land-type fencing and revegetation around this gully and dam on 'Helm View', as the trees have grown so fast. Within four years, a saline, eroding drainage line and potentially salty dam has been converted to an asset, for the farmer and for the local ecosystem.

River. Peter ran a grazing enterprise of Merino sheep and Hereford cattle.

'Benowrie' is well timbered with single old Red Gums dotted across the landscape, but Peter had noticed an increasing number of dead trees each year before his involvement in the Potter Farmland Plan, and tree decline is significant in the area. There is also some minor gully erosion and gully salinisation. 'Benowrie' was included in the Potter Farmland Plan because it is adjacent to the Casterton-Edenhope Road. It also offered an opportunity to demonstrate the potential for rehabilitating mature or over-mature Red Gum (*Eucalyptus camaldulensis*) woodland through encouragement of natural regeneration, and establishment of clumps incorporating understorey species.

'Satimer'

'Satimer' is situated on Satimer Road, 30 kilometres north-east of Casterton and 36 kilometres north-west of Coleraine, and has been in the Speirs family for most of this century. The 1880-hectare farm is presently owned by Roy Speirs and his sons, Bill and Jack, who together with their wives, Prue and Suzie, operate a Merino sheep stud and Simmental cattle enterprise, mixed with some cropping and commercial hay production. Satimer Road runs through the middle of the property.

The 'Satimer' landscape is typical of the Wando Vale/Nareen area, with some high plateau land, steep escarpments and deep gullies and some more gentle slopes. The central portion of the farm was thoroughly cleared many years ago and is now very bare, with emerging salting in lower areas. In fact it may have been cleared by my great-great-grandfather, Alexander Clay-hills Cameron, who, I discovered a year into the project, had settled on 'Satimer' from 1862-74. The northern section of the farm is dissected by the Wando River, and the southern end of the property is dissected by Davidsons Creek, and Corea Creek. Gully erosion, hillside slumps and soaks have been a major problem along these creeks and their tributaries for many years, but the Speirs family has had a long involvement with soil conservation works. The Soil Conservation Authority started erosion control works on 'Satimer' in the late 1950s and these works continued into the 1980s.

'Satimer' joins 'Barnaby' to the north-east, and 'Gheringap' to the south, so the three farms make a continuous demonstration along Satimer Road. There are two significant areas of remnant native woodland on Satimer. One forms a continuous forest along the boundary with 'Barnaby', and the other, which is also a Manna Gum/Snow Gum association, is on the southern boundary, next to 'Gheringap'.

MELVILLE FOREST

'Pine Grove'

John and Helen Diprose own and operate 'Pine Grove', a 1000–hectare farm 10 kilometres south-west of Cavendish. Like most farms in the Melville Forest-Bulart-Cavendish district, 'Pine Grove' is undulating, with duplex soils—grey loams over yellow clay and some 'buckshot' gravel banks—and a sparse cover of huge mature or over-mature Red Gums (*Eucalyptus camaldulensis*), many in advanced stages of decline. Again, like most farms in the district, 'Pine Grove' has some expanding is salt-affected flats along the creeks and some very minor erosion of creek banks, and it is completely surrounded by roads (sealed on two sides), with a gravel road through the centre of the farm.

'Pine Grove' is an aggregation of four old 'soldier-settler' blocks (the most recent purchased in 1980) and thus has a variety of fencing layouts and fence quality. Water for stock is provided by dams and three bores which provide water of poor quality. John has been engaged on an active programme to increase wool production through incorporation of more productive blood lines and pasture improvement. The whole farm plan prepared during the Potter Farmland Plan rationalises farm layout (within the constraints set by roads) and water supply and builds on the programme of pasture and fencing improvements started by John.

'Wyola'

Gavin and Vicky Lewis operate 'Wyola', a 180–hectare block which adjoins Pine Grove to the south-west, and also play a major role in running 'Anchorage', owned by Gavin's parents, Ted and Joan. The landscape is very similar to 'Pine Grove', although salt flats are more extensive and 13.6 per cent of the property is salt-affected. When Gavin and Vicky took over 'Wyola' it was unimproved, with a few large paddocks fenced on the square and annual pastures of low quality. Water supply was also poor, relying mainly on Dundas Creek, which is marginal for stock.

'Warooka'

John and Joan Lyons have owned the 580–hectare property 'Warooka', two kilometres south-west of Mt Dundas, since 1967 and they also lease another 232 hectares on the other side of the mountain. 'Warooka' has similar soils and topography to the other Melville Forest farms, with the addition of some small patches of black soil on creek flats and a beautiful, relatively healthy and dense woodland of old Red Gums extending over most of the northern and eastern portions of the farm.

John prepared a whole farm plan in the late 1960s, fencing creek flats and heavy soils into separate land units. He immediately began a pasture improvement programme based on direct drilling, and subdivided the property with a laneway in 1974. John and Joan run a very efficient operation, doing most work except shearing and crutching themselves. John aims to renovate 10 per cent of his pastures annually, and manages pastures carefully with a boom spray and livestock to reduce weed content. He also grows some pastures for seed production and Persian clover for high-protein hay.

'Helm View'

Colin Milne and his sons John, Bruce and Andrew and their families own 'Helm View', which is a 524-hectare block 1 kilometre south-east of Mt Dundas, across the Dundas Gap Road from 'Warooka'. The Milne family also owns an additional 200-hectare block at Melville Forest and 400 hectares near Penshurst. The Milnes run a Hereford cattle stud and medium-wool Merino sheep. 'Helm View' is a very bare block, stark and exposed, with the dominant feature of the landscape being thousands of large dead Red Gums, ringbarked in the 1930s.

'Helm View' was very run down when the present family partnership took over in 1965, with virtually useless fences, unimproved pastures, no shelter and poor water supplies. Bruce Milne prepared a fencing plan in 1966 and, over the next nineteen years, the Milnes carried out an intensive programme of stock breeding, fencing, construction of sheds and yards, pasture improvement and the development of a water-supply system based on a 15 000-cubic-metre dam, high on the colluvial slopes of Mt Dundas. During this period the Milnes did everything themselves and supplemented their income with shearing off-farm.

Two drainage lines begin on 'Helm View' yet even at the top of the catchment, salinity is evident and spreading, with 7–8 per cent of the property affected. By 1985 erosion problems were minimal, due largely to land-type fencing, but when the Potter Farmland Plan began, apart from some Red Gum, Yellow Box (*Eucalyptus melliodora*) and Drooping Sheoak (*Allocasuarina verticillata*) remnants on the north-eastern edge of the farm, tree decline was total.

'Willandra'

Peter and Julie Waldron own 'Willandra', a 356-hectare property about five kilometres south of 'Helm View'. They run medium-wool Merinos and a few Angus cattle. 'Willandra' has been in Peter's family for many years, and Peter helped his father to develop the farm, grubbing out stumps, building fences and sowing new pastures from the 1950s to the 1970s, while also working as a wood merchant and a shearer. 'Willandra' is a typical Melville Forest block. It is gently undulating, the few remaining old Red Gums are senescent and the soils are grey loams over clay. Saline gullies comprise about 5 per cent of the farm and dams in gullies are now too salty for stock.

GLENTHOMPSON

'Ballantrae'

Doug and Helen Heard, their son Leigh and his wife, Phillipa, operate 'Ballantrae', a 1300–hectare property on the Glenelg Highway, two kilometres east of Glenthompson. 'Ballantrae' has been in the Heard family since 1934, with additions in the 1950s and 1987. They run a self-replacing medium-wool Merino flock, a small cattle herd and crop 5–10 per cent of the farm each year.

'Ballantrae' is a farm of thoroughly cleared, gently rolling hills, sitting astride an east-west divide, with the northern half of the property draining into the Wannon and the southern half draining to the Hopkins River. The soils are generally grey sandy loams over yellow clays and the only vegetation on the property in 1985, apart from improved perennial pastures, was a small patch of remnant Red Gums, some mature pine and cypress corner clumps and plantations and an avenue of Sugar Gums along the 'Ballantrae' driveway. The highly dispersible yellow clay subsoils have eroded in some gullies, and there is a small saline flat along one creek. Doug Heard has had an active involvement in soil conservation works for more than thirty years and has controlled the gully erosion problems on 'Ballantrae' by land-type fencing, diversion banks, trickle pipes and concrete-drop structures. Like many farms in the Glenthompson area, 'Ballantrae' is also vulnerable to hill-top wind erosion in dry years. Doug and Leigh have been resowing pastures for the last ten years and had begun establishing a laneway system before their involvement in the Potter Farmland Plan.

'Cherrymount'

Dave and Prue McCulloch's farm, 'Cherrymount', is an 800–hectare farm nine kilometres south of Glenthompson on the Caramut Road. Dave farms about 80 per cent sheep and 20 per cent cattle. 'Cherrymount' is high in the catchment of the middle reaches of the Hopkins, shaped in a long narrow rectangle running along a north-south divide. There is one small patch of remnant Messmate Stringybark (*Eucalyptus obliqua*) and the rest of the sandy loam soils are bare of trees. A creek which flows from the Glenthompson-Caramut Road through the south-west corner of the property is severely eroded, as are drainage lines which flow to the north from 'Cherrymount' and which have made Cherrymount Lane impassable. Gullies which drain from 'Cherrymount' to the south into McDonalds' and to the east into Levinsons' are extremely saline, with expanding, unproductive salt flats. 'Cherrymount' was included in the Potter Farmland Plan to complete treatment of the sub-catchment to the north of McDonald's.

'Daryn Rise'

David McDonald and his wife, Robyn, farm 'Daryn Rise', a 255-hectare (one square mile) block on Yarrack Road, 13 kilometres south of Glenthompson. The block is completely devoid of native trees, apart from one small straggly clump of Messmate (*Eucalyptus obliqua*). The property is dominated by two drainage lines which run north-south out of 'Cherrymount'. The head of one of the gullies was actively eroding in 1985 when the Potter Farmland Plan began. Both gullies are very saline, with salt levels as high as sea water in autumn and small saline soaks breaking out on higher slopes. Saline areas now make up 15 per cent of the farm.

David and his father, Russ, bought the block in 1978 in unimproved condition, poorly subdivided, with unimproved annual grasses, obvious salinity and erosion and not much in the way of buildings, yards, fences or water supply. The market value reflected this condition, giving Russ and David an ideal opportunity to develop the farm. They worked to sow improved pastures, to build a new woolshed and a house for David and Robyn and to initiate a laneway system, before their involvement in the project.

'Fernleigh'

Stuart and Lavinia Cuming operate 'Fernleigh', a 1712-hectare property on the Glenthompson-Caramut Road, 15 kilometres south of Glenthompson. Stuart runs a Comeback sheep stud, with some cattle, and about 10 per cent of the property is in crop each year. The distinguishing features of 'Fernleigh' are its soil, which is predominantly a deep sand or sandy loam, and the remnant patches of woodland dominated by Manna Gum (*Eucalyptus viminalis*) with an understorey of Blackwood (*Acacia melanoxylon*), Golden Wattle (*Acacia pycnantha*), Black Wattle (*Acacia mearnsii*), Sweet Bursaria (*Bursaria spinosa*) and Silver Banksia (*Banksia marginata*). Tree decline among these remnants is extreme, the area of saline drainage lines is expanding and wind erosion on sandy banks in dry seasons is always a threat, of which Stuart is well aware.

Like 'Nareeb Nareeb', 'Fernleigh' was extensively burnt in the grassfire of 12 February 1977 and 55 kilometres of fencing was renewed in a major redevelopment of the farm. Pastures are continually renovated on a seven-year rotation which includes cereal and legume crops and regular additions of trace elements and fertilisers to the sandy soil.

'Nareeb Nareeb'

'Nareeb Nareeb' has been in the Beggs family since the 1870s. Hugh and Frankie Beggs run the 3060–hectare property as a fine-wool Merino stud, complemented by a herd of Hereford cattle. 'Nareeb Nareeb' is situated about 60 kilometres east of Hamilton and 20 kilometres south of Glenthompson. The Glenthompson-Caramut Road runs through the property from north to south and the Hamilton-Chatsworth Road runs along its southern boundary. Soils are white sands in the northern paddocks adjoining 'Fernleigh', grey podsols in the central portions and grey basaltic loams and clays in the south. 'Nareeb Nareeb' is undulating, with mainly bare slopes apart from a scattered woodland of old Red Gums in the central and eastern portion and mature native and coniferous clumps and shelter belts established by Hugh's father. Most of the drainage lines on 'Nareeb Nareeb' are saline and the salt flats are extensive, with some minor instances of gully erosion. Many dams, especially the larger dams in the gullies, have gone saline in the last twenty years and are now useless for stock.

Awareness of the threat of dry-land salting in the area emerged long before the Potter Farmland Plan. Hugh Beggs has planned and supervised an active programme of fencing and renewal of perennial pastures and has a long-term policy of tree planting, both in plantations and individuals. Seventy-five per cent of 'Nareeb Nareeb' was burnt in the 1977 fires and the property suffered from wind erosion in the drought of 1982–3.

'Reedy Creek'

Barry and Ione Levinson purchased 'Reedy Creek' in 1981 after moving from Kojonup in south-west Western Australia. They were attracted to the 1380–hectare property by its rolling hills and creeks which wind their way through the farm. The 'Reedy Creek' property is on Phillips Lane, 10 kilometres south-east of Glenthompson. It is generally bare, with a few areas of remnant Red Gum woodland, particularly along the creeks, which are also lined with indigenous understorey shrubs such as Bushy Needlewood (*Hakea sericea*), Hedge Wattle (*Acacia paradoxa*), Blackwood (*Acacia melanoxylon*), Black Wattle (*Acacia mearnsii*), Drooping Sheoak (*Allocasuarina verticillata*), and Slaty Sheoak (*Allocasuarina muellereana*). Soils are predominantly sandy loams, except for an exposed ridge of Cambrian greenstone which runs in a north-south belt on the eastern side of the property.

Dry-land salinity is expanding in one flat in the south-west corner of the property. From the moment Barry and Ione took over 'Reedy Creek', they have had a long-term goal of fencing out all the creeks to encourage regeneration and control the active gully erosion which was evident, and also to fence off all the hill tops and plant them to trees. The latter goal was reinforced during the 1982–3 drought, when wind erosion on the sandy hill tops was severe.

Demonstration Farm Summary							
Farm	Size (ha)	Years owned	Rain (mm)	Labour units	Sheep % d.s.e.[1]	Cattle % d.s.e.[1]	Crop % area
Melville Forest							
'Helm View'	524[2]	44	650	3.5	62	38	
'Pine Grove'	1003	51	675	2	97	3	5
'Warooka'	580[2]	21	675	1.5	100		
'Willandra'	356	58	650	1	80	17	3
'Wyola'	180[2]	26	700	1	100		
Wando Vale							
'Barnaby'	650	13	750	1.1	80	20	
'Benowrie'	972	24	710	1.75	85	15	1
'Gheringap'	560	10	725	1	100		
'Satimer'	1880	55	800	4	98	2	3
Glenthompson							
'Ballantrae'	1300	56	650	2	85	15	7.5
'Cherrymount'	800	53	650	2	80	20	7.5
'Daryn Rise'	255[2]	13	700	1	100		
'Fernleigh'	1712	55	670	3.5	80	10	10
'Nareeb Nareeb'	3060	81	575	7	80	20	2
'Reedy Creek'	1380	10	700	2	100		5

[1] d.s.e. = dry sheep equivalents
[2] Indicates the main property only—other land is owned or leased elsewhere.
'Years owned' refers to the number of years the farm has been in the family to 1991.

3

Whole Farm Planning

● THE POTTER APPROACH IN CONTEXT

FARM PLANNING is not new in Australian agriculture. Various approaches to farm planning have been used, usually to address 'problems', such as controlling soil erosion, or ensuring efficient water use.

The term 'whole farm planning' has been in common use in irrigation districts to refer to the process of redesigning the supply, delivery and removal of irrigation water from a farm. In these cases, the term indicates that planning is carried out over a whole farm, as opposed to considering each irrigation bay or each paddock in isolation. Many plans for irrigation properties encompass more than just the movement of water—they consider stock movement and access lanes as well as physical improvements such as buildings and yards, but the primary focus of the plan is towards the engineering works required to ensure the most efficient delivery of water to the whole farm.

Soil conservation agencies have used whole farm plans to outline soil conservation works over a whole farm. Until the late 1980s in Western Australia, New South Wales, South Australia and Queensland, the concentration has been on soil conservation earthworks such as contour banks, complemented by changes to farm layout based on land classes, adoption of minimum tillage, improved crop rotations and water supply improvements.

In Victoria, the emphasis has been slightly different since perennial pastures are used rather than contour banks in most areas, so that plans are less dominated by the siting of earthworks, although the approach taken in cropping areas is similar to that of other states. Victorian soil conservation plans usually outline changes to farm layout based on land classes, pasture and water supply improvements and specific erosion control structures such as spillways, trickle pipes, groynes and tree planting for gully stabilisation. However, the primary focus of all of these plans has been to minimise soil erosion and, in some cases, salinity.

The essence of the approach applied on the Potter Farmland Plan demonstration farms is that land-degradation 'problems'—such as erosion, salinity and the largely ignored issue of tree decline—are *symptoms* of inappropriate farmland management. There is no point in just focusing on the symptom without tackling the cause of the problem. That is easy to say, but for a farmer or a soil conservationist confronted by an erosion gully chewing its way across the landscape, it is difficult not to focus on the gully itself. As the old saying goes, 'when you're up to your arse in crocodiles it is easy to forget to drain the swamp!'

The philosophy we started with at Hamilton was that such problems can only be effectively addressed if farm design and management incorporates an understanding of the ecology of the land. Unless we restore ecological stability to farmland, we will always have 'problems', and the productive capacity of the land will continue to decline. Taking the simple gully erosion case above, the long-term solution is to develop a farming system of crops, pastures, trees, livestock and undisturbed areas which ensures optimum water use, controls excess run-off, provides shelter and optimises soil fertility and stability. In the short term, of course, the active gully erosion would have to be controlled, probably using a combination of fencing out, earthworks and revegetation to restore some ecological diversity and stability to the drainage line.

Given the financial pressures faced by farmers, however, an ecological balance is only likely to be achieved on most farms if the changes required complement or enhance farm production, both in the short and the long term. Our firm belief at the start of the project, which still holds today, is that it is almost always possible to improve the stability of a farming system without great expense, given a liberal dose of lateral thinking and sufficient commitment to finding a better way. Whole farm plans provide a framework for blending ecological principles into agricultural systems. The extent of land degradation in Australia is ample evidence that farming systems which only focus on maximising agricultural outputs or the financial bottom line are untenable in the long term. Equally, it should be recognised that most Australian farmland consists of severely modified ecosystems and it is not always possible, and rarely practical, to return them to their original state. Farms which are ecologically sound but which do not produce enough to sustain a livelihood are at the other extreme from the exploitative, degrading farm and are also inappropriate.

That is not to say that every area is capable of producing enough to sustain a livelihood in an ecologically stable manner. There are many areas of Australia, particularly in the arid zone, where the inherent productivity of the natural ecosystems is so low and climatic variations so extreme that there is very little surplus for man to harvest, however sensitively, particularly given the feral animals and weeds which have been introduced over the last two centuries. Whole farm planning forces an examination of the capacity of the land and exposes instances where land is used beyond its capability.

The whole farm plans developed on the Potter Farmland Plan demonstration farms map the natural and man-made resources on the farm, analyse those resources and the present and projected needs of the farm enterprise and outline farm improvement programmes aimed at achieving more sustainable and profitable use of the land. The whole farm plan considers the whole farm and all its planning needs, not just a single development or problem, and provides a framework for analysing and integrating personal, business and land-development goals and planning for change, rather than reacting to it.

● PRINCIPLES

The principles which were the basis of the whole farm planning approach applied on the Potter farms are summarised below. These principles were established before the project was initiated at Hamilton. They build on land-classing systems which have been used by soil conservation agencies in Australia and America for many years, incorporating ecological principles drawn from the work of the Garden State Committee (in particular John Jack, Carrick Chambers and Bill Middleton), and the emerging sustainable agriculture movement.

- Farms should be subdivided into homogeneous land units based on natural features such as soil type, slope, drainage and vegetation, rather than on arbitrary boundaries made by people.
- Each land unit should be managed according to its potential and its limitations (potential to degrade), with an understanding of the ecological processes in operation both within the farm boundary and over the land system and catchment in which the farm is located.

- Farm improvements such as water supply, drainage, crop and pasture improvements, access roads and revegetation should not be looked at in isolation, but rather integrated into a plan which considers the farm as a whole, not a collection of discrete parcels of land.
- Farm management should aim to incorporate the elements of existing natural systems which convey the stability, resilience and ability to recover from disturbance which characterise a sustainable system. These elements include structural and species diversity within populations of plants, animals and micro-organisms, efficiency in the use of energy and resources, optimal turnover and recycling of matter and nutrients, efficient energy flow, adaptation of crop and animal components to the environment and conservation of the renewable resource.

Whilst the principles above were clear from the beginning, the whole farm planning process itself evolved with the knowledge and experience of the staff and participating farmers as the Potter Farmland Plan proceeded.

● RESOURCES USED IN PREPARING PLANS

We decided at the Creswick consultation that the demonstration farms would be planned using information and technologies available to the average farmer. We had briefly investigated the possibility of commissioning detailed surveys of each farm—by hydrologists to examine water-table issues; soil scientists to map soil boundaries, structures, textures and fertility levels; pasture experts to look at species composition and plant nutrient status; and botanists to look at remnant vegetation—but these surveys were expensive and they would have taken at least a year to complete for the fifteen farms. Such detailed resource inventories are beyond the capacity of most farmers and anyone wishing to follow the example of the Potter farms would have been constrained if the planning process relied on an expensive initial analysis for the farm by outside experts.

The procedure used in planning the farms was as follows:

1 We obtained photopositives of aerial photographs of each of the demonstration farms from the then aerial photo section of the Lands Department (now part of the Department of Property and Services) in Treasury Place, Melbourne. The aerial photo section provided the initial photopositives, worth $150-200 each, free of charge as a contribution to the project. Aerial photographs can be obtained in most states through departments of lands or soil conservation and most states are now streamlining these services to cater for the demand for farm plans.

2 The farm was inspected by project staff and the demonstration farmers over several days, to discuss the land, its history and problems, the type of enterprise, and the resources available to carry out the plan. Any developments or particular issues that the demonstration farmer wanted the plan to address were discussed, until project staff had a good 'feel' for the property.

3 Project staff prepared an initial draft of the plan to put forward ideas for discussion and introduce the demonstration farmers to the principles mentioned above. These early drafts were prepared by John Marriott and me, helped by Bill Middleton and Bill Sharp, with occasional input from Peter Dixon (DCE—water supply, erosion and salinity) and Peter Schroder (DARA—pastures).

4 The first draft was discussed with the farmers, who were encouraged to make alterations and additions. It was left with the farmers for several weeks, to enable them to digest the implications of the plan, to discuss it with family members and to add detail as their thoughts on the basic framework became clearer. This 'kitchen table' farm planning was very important, as it helped demystify the process and got the farmers involved in preparing their own plans.

5 When everyone was happy with the first draft, the high priority elements of the plan were identified, and a works schedule for the first year (1985) was drawn up.

6 As the first elements of the plans were put into practice, farmers were encouraged to evaluate their plans continually and to decide which portions of their plan should be high priorities for implementation in subsequent years of the project.

7 Most plans went through a continual evolution, with increasingly more input from the farmers and less from project staff.

The initial draft plans were not static documents. It soon became obvious, for instance, that to prepare a whole farm plan properly, it is necessary to know what the land is like at all times of year. Farm layouts on several plans drafted in summer or autumn were revised after observing soil conditions in winter and seeing which fences and gateways were subject to most pressure. Small but critical details such as these were known intimately by the farmers, but were very difficult for project staff or other outsiders to pick up in a short time. By contrast, it is often difficult for someone who has managed a farm in a certain way for many years to rethink the design of the farm, as fences, tracks and water supplies tend to become fixtures in the farmer's mind. Usually the longer a farmer has been on the farm, the harder it is to initiate a whole farm plan.

With the benefit of hindsight and the opportunity to see farm and catchment planning approaches in other areas which have developed since 1985 (particularly during the rapid growth of landcare in 1989 and 1990), if I had the opportunity again I would put a great deal more emphasis on farm inventory than we did in the rush to get the Potter plans started. This would not involve commissioning surveys by experts, it would involve much greater encouragement of the farmers to gather their own resource information, consulting specialists where necessary, examining in detail the physical resources of the farm and matching this information against financial and enterprise information. The primary purpose would be to assist the farmers to document what they know about their land, and even more importantly what they do not know, as a spur to find more information, and the significance of this information. It may delay the start of works, but it would ensure that the works are based on the best information available and, even more importantly, it would expose the farmers to new information and encourage them to seek new sources of information very early in the process.

The decision made at Creswick to base our resource inventories on the information available to the average farmer, so that other farmers would not feel inhibited by the process, was sound. But I now believe that we have to change the prevailing thinking about types of resource information which can and should be gathered by farmers. Some farmers are making use of technology such as infra-red aerial photography to monitor crop vigour, ground-based and aerial electromagnetic salinity hazard mapping, neutron probes for soil moisture status, soil and plant fertility testing and so on. The impact on farmers of a flight over their property to see it from a new perspective, and of digging a few soil profiles and having them interpreted by someone capable of explaining what is happening beneath the surface, can be profound, but few farmers get these opportunities. We need to make such technology far more accessible to the average farmer and I believe it is now appropriate to expect farm plans to be based on this type of information, to complement the detailed insights gained by the farmer in living and working close to the land.

It was fascinating for me to hear Bruce Milne give the keynote address to the Victorian LandCare Conference in Wangaratta in July 1990, during which he dared to question the often-made statement that 'farmers are the true conservationists and know their land best'. Bruce related a discussion with an ecologist, who asked him about the frogs (species and populations) on 'Helm View', and about the native grass, herb and insect communities which would have been there prior to settlement, to try and get a picture of the ecological stability of the existing system. Bruce was unable to answer the questions, though he could see their relevance to understanding more about the long-term trends in water quality, species diversity and the interactions between plants, insects, animals, soil and water. He has always walked around the farm with his eyes open and his mind in gear, but now he sees and thinks about more than stock and pastures and budgets. We could have hastened this process if we had put more effort into a comprehensive farm analysis and inventory at the start of the project.

Once the farmers found it possible to accept the idea of radical changes to the existing farm layout or enterprise the whole farm plans developed at a pace and level of detail which outsiders could not have contemplated. For example, Peter Waldron, after careful consideration of early draft plans, designed a laneway which ran from his woolshed almost to the south-east corner of his property, diagonally across a chequerboard layout of quite sound fences on a north-south, east-west grid which Peter and his father had constructed. The laneway meant that several fences had to be removed, but it resulted in a dramatic improvement to stock management around the farm and to farm management generally. With the confidence this single development gave him, Peter took over the full responsibility for his plan.

Whole farm plans evolve and the planning process should never be finished.

● THE WHOLE FARM PLANNING PROCESS

FARM ANALYSIS

The Base Map

Aerial photographs were used as base maps, since they show both natural features and many farm improvements in detail, which saves a lot of surveying, measuring and drafting. On the demonstration farms the photographs were obtained from the Department of Property and Services as a positive enlarged to a scale of 1:5000 (farms less than 750 hectares) or 1:10 000 (larger farms) on 'chronoflex', a durable clear film. The advantage of this format is that multiple paper copies of the photograph can easily be copied from the master film, for about $1 each, on a dyeline machine at the local municipal office. It was two to three times more expensive than simply getting an enlarged photo, and the paper copies were not quite as sharp as the original photo, but the ability to produce a large number of effectively disposable photos, from, one master copy was a major advantage. The chronoflex format was particularly suited to the demands of the Potter

Farmland Plan, where it was essential to have several copies of each plan—several for the farmer, one for recording works details, one for planning future developments, and one for use on farm inspection tours and displays. It allowed for plans to be used as working drawings in several farm buildings and vehicles at once, and it allowed successive or alternative drafts of the plan to be prepared quickly, without worrying about possible damage to the original.

I have since seen a terrific cheap alternative to the chronoflex positive for making a number of copies of an aerial photograph base map for farm planning. Mike Clarke, a project officer with the Western Australian Department of Agriculture at Geraldton, assists several local Land Conservation District Committees with farm and catchment planning and has developed a very simple but effective system. It simply involves borrowing survey format aerial photographs (from the Department of Land Administration) and photocopying them on a laser photocopier, available at most large regional centres. These photocopiers can enlarge 1:20 000 or 1:25 000 photos, say, two times or five times to get very good copies of the desired area at 1:10 000 or 1:5000 scale (or any other desired scale). These can then be cut and pasted to make maps of larger areas and laminated to make a robust base which can be drawn on with coloured pens, for a cost in 1990 of about twenty dollars.

John Marriott has developed a further refinement since he began his farm planning consultancy in 1989, which exemplifies his belief in keeping the planning process simple and accessible to the farmer. John fixes strips of magnetised rubber (obtained from businesses who make signs) to the back of the aerial photograph base map. This enables the plan to be fixed easily to any metallic surface (such as the fridge door or the car bonnet), keeping it flat and preventing it from flapping in the wind. Strips of metal can be placed on the walls of the office and sheds (the back of the toilet door is a strategic place for reflection on future directions) so that the plan(s) can be easily moved and displayed to best advantage.

The Potter plans (see colour maps between pages 88-9) consisted of two clear plastic overlays over the aerial photo base map. Natural features of the farm were depicted on the base map, the existing layout and man-made improvements were depicted on one overlay (overlay A), and planned future developments were outlined on a second overlay (B).

The clear plastic overlay with coloured pens on an aerial photograph is a simple system which is a very effective planning tool for farmers, land-use planners and agricultural advisers, but the potential of computer technology to be used in drafting and aggregating farm plans is being realised in a number of places, particularly in Western Australia and New South

• A typical Glenthompson landscape today. This area was extensively cleared and now consists of bare, windswept, rolling hills dissected by creeks and gullies which are often salty and eroded.

• The Melville Forest landscape from Mt Dundas looking over 'Helm View', the Milne demonstration farm in the middle ground. A scattered, open, savannah woodland of Red Gum remains, with some shrub species on roadsides, but this landscape has been heavily modified since settlement. Clearing, ringbarking, new pastures and grazing have altered the hydrologic balance, causing salinity which is often accompanied by erosion in low areas.

• The best place for whole-farm planning is in the paddock. The author (left) and David McCulloch discuss suggestions on the whole-farm plan for 'Cherrymount' in February 1985. The plans for the demonstration farms evolved through many drafts, with fewer and fewer inputs from project staff and more and more from the farmers.

• Before and after: this 'Willandra' laneway through a saline gully was just being established in 1985 when the upper photograph was taken. Five years later (lower photograph), the gully has been stabilised, the lane is sheltered and protected, land-type fencing according to the whole-farm plan enables fine tuning of grazing and Peter Waldron has a more flexible and sustainable system.

Wales. Resource information such as geology and soils, water tables, land classes, vegetation, wildlife species or even population densities can be digitised and stored in a computer. In the right hands and with the right equipment, this enables integration of various types of information, quick retrieval, alteration or drafting of plans and most importantly, the ability to aggregate large-scale plans such as farm plans, into smaller scale regional or catchment plans.

These systems, usually called 'geographic information systems' (GIS) will undoubtedly make a great contribution to rural land use planning over the next decade. But it is essential to remember that, no matter how impressive the digitised images are on the screen, they are only as good as the information which has been put into the system. In the case of farm plans, the farmer should be intimately involved in compiling this information, both to get local knowledge and to develop a sense of *ownership*. It is also important that farmers have constant access to their plans to enable continual refinement and evolution of the plan to occur. Ownership is the critical point, and it is much more difficult to achieve with a planning process which relies on technology which few farmers understand and equipment few farmers have. While this situation will change as farmers adopt computer technology, the fundamental importance of ensuring that plans reflect detailed local knowledge will remain.

This book concentrates on the whole farm planning approach which evolved on the Potter Farmland Plan demonstration farms from 1984 to 1988. However, the principles of the Potter approach have been applied in other areas in the late 1980s, and some innovative farm-planning processes have been developed, particularly involving groups of farmers planning on a catchment scale. To illustrate some of the new directions in planning for more sustainable farming I will digress briefly, moving forward five years to a Western Australian case study.

One of the best approaches to farm and catchment planning I have seen is being used by a private consultant, Viv Read, working with Land Conservation District Committees on National Soil Conservation Program projects in the wheatbelt of south-west Western Australia. Viv works with groups of farmers who wish to plan their properties to improve sustainable productivity while redressing and preventing land degradation, which demands that plans for individual farms must form a coherent approach on the catchment scale.

The approach Viv has developed with the farmers is simple:

• Viv and as many as possible of the farmers in the catchment (or sub-catchment) get together to discuss the strengths and weaknesses of the existing farming system, the causes and extent of land degradation problems and the elements of a sustainable farming system for the area. A feature of this process which appeals to me is the use of soil peels to help the farmers to get a common understanding of the different local land types and their characteristics. Viv and the farmers inspect and describe representative soil profiles in the field, using a sheet of latex-coated cheesecloth to preserve a 'slice' of the profile for future reference. They then agree on a common name for and the capabilities of each land type, so that when one farmer talks about a suitable land use for 'deep yellow sands' or ' sand over gravel', it means the same for all participants.

- The farmers then work individually on their whole farm plans, assisted by Viv, who considers his input as 'the voice of the catchment', ensuring that the individual farm plans aggregate into a coherent approach for the whole catchment. Alternative enterprises such as wildflowers, fodder trees and nut crops are considered and land conservation earthworks are coordinated across the catchment, involving local government authorities where necessary. Farmers continue to work on these hand-drawn plans with Viv advising on farm layout, land capability and alternative land uses.

- When the farmers are well-advanced with their individual plans, Viv prepares digitised plans for each farm and the catchment as a whole.

By this stage in the process the farmers are fully involved in their farm planning and committed to the works proposed. There is a healthy balance between high-technology computer digitising and plotting of plans and the 'hands-on' work in the paddock and scribbling on rough drafts, and between farmers planning their own patch and the consultant advising from a catchment perspective. The farmers' commitment towards the works proposed is an indicator of the ownership they feel for the process. The value of the consultant is illustrated by the number of new ideas incorporated into farm plans and the cohesion of the farm plans when aggregated to the catchment scale.

There are two ways to measure the worth of a farm plan—by the changes it engenders inside the farmer's head and by the resulting changes on the ground. The method of compiling, representing and storing information does not matter, providing it improves farmers' understanding of their land and of their impact on it, facilitates better planning and decision-making about land use and resource allocation and leads to more sustainable use of the land.

Natural Land Units

Using the aerial photo as a base map, we began on the Potter Farmland Plan farms in early 1985, marking in natural features such as watercourses, drainage lines, ridge lines, vegetation types, soil types (as perceived from the farmer's experience and visual assessment) and catchment boundaries, relying a great deal on the farmers.

We tried to ignore existing improvements such as fences and roads, concentrating on the land itself, as existing fences, drains and tracks can often subconsciously influence attempts to subdivide the land into land units. The natural boundaries between different types of land on the property were then outlined. These boundaries were based on soil type, aspect (for example, north facing or south facing), slope, drainage and vegetation.

In some ways this process was similar to the land-classing systems used by soil conservation services for many years, but other factors were also considered. Formal land-classing systems in Australia divide land according to its suitability for cultivation, which takes into account soil type, slope and drainage. These systems are an ideal basis for a plan which has as its primary focus the control of soil erosion. However, when planning a farm to achieve broad production and ecological goals, there are factors other than the suitability for cultivation which should be considered. One half of a paddock may be sheltered compared with another, for example. The line between these land types is a logical subdivision provided each is a viable management unit.

We relied on the farmers' knowledge of their land to suggest these natural boundaries.

Land Capability

We then considered the use to which each type of land was most suited within the farm enterprise and also the potential for different land uses. This was an important step, as it forced farmers to re-examine the potential and the limitations of each land unit, a prerequisite for designing new layouts and water supply and vegetation strategies.

Viv Read has refined the order in which these issues are considered in the planning process, which I believe is an improvement. After identifying resources (including land units), then land management hazards, Viv and the farmers in the catchment group consider sustainable agricultural systems for each land unit before preparing a draft plan. In other words, the first question asked before designing a new layout should be: 'What is a sustainable land use for this land unit?' Obviously this will be very difficult to answer, but it does throw up the questions which land users need to be asking themselves if their farm plans are to be of lasting benefit. The preparation of draft plans includes broad knowledge and decision making about agricultural systems. Once again with the benefit of hindsight, this refinement would have improved the farm planning process we developed on the demonstration farms.

In the main we refined existing farming systems in the direction of sustainablility, rather than looking for completely new systems, although alternative enterprises were discussed occasionally with several farmers. Ernesto Sirolli, after visiting the farms in 1987, asked: *'These are very fine farms, but do you really think you could still be grazing sheep here in one thousand years?'* I had to say no. It is a challenging question which we should have been asking at the start of the project, and which all land users should ask themselves from time to time, in order to question the 'because-we've-always-done-it' habit and to stay open to new possibilities.

We used land-systems information as a guide to land capability at the district level. However at the individual farm level we relied mainly on our own experience and the farmer's local knowledge, with occasional specialist inputs from local staff from CFL and DARA.

Although there may be very detailed written information available, it is important that landholders attempt to identify their own natural land units before seeking help. With both the PFP farms and subsequent short courses, this revealed how much landholders did *not* know about their land, as well as what else was known. Participants in the Potter Farmland Plan say that after going through the whole farm planning process, they now have a much deeper knowledge and appreciation of their land, its qualities, limitations and potential.

In most instances on the demonstration farms, the land capability assessment was not an explicit step in the planning process, but implicit in the thinking behind revegetation and water-supply strategies. It was more a by-product of the planning process than a prerequisite for plan development. As the farmers increased their knowledge of their land, they began to question their own use of different land units more carefully. For example Barry Levinson on 'Reedy Creek' decided to move away from broad-acre cereal cropping on his land as he learnt more about the relative degrees of water use of crops and perennial pastures and became more concerned about the prospect of dry-land salinity.

We also examined the impact of past land use on the land and the 'sensitive' areas of the farm, such as eroding, saline or overgrazed areas and places where the quality of pastures had declined. These areas were usually well known to the farmers and required particular attention in the planning process.

We also marked on the base map other areas which were difficult to manage, for example steep, rocky, waterlogged or infertile areas. Possible hazards, such as areas which are prone to fire, erosion, flooding, frost, vermin or disease, were also noted on the base map, as were the direction of critical winds for stock and crop shelter and fire protection.

Although farmers knew which parts of their farms were prone to some of the hazards mentioned above in broad terms, the act of drawing a line on an overlay to delineate a particular area often required further checking in the paddock, which exposed in more detail any gaps in knowledge. This often stimulated a reappraisal of some of the underlying imbalances of which the 'problems' were merely symptoms. This exploration of the key land-use issues early in the planning process was very important in developing the thinking of the farmers and project staff as planning evolved.

The first overlay (overlay A) was prepared by attaching a sheet of clear plastic to the base map, using transparent tape along one edge only, so the overlay could be flipped aside easily. On overlay A, the demonstration farmers were asked to use a waterfast pencil or coloured pen to outline the existing layout of the property, marking in fencelines (and gates), paddock names and areas, roads, laneways, yards, buildings and water supply facilities. The farmers also listed the present use or uses for each paddock, in some cases using a key to indicate the age and quality of different fences. Planning constraints such as land tenure—government or private, leasehold or freehold—or boundaries of individual holdings within the family or farm partnership were marked on this overlay, as were utility easements such as power lines, pipelines or communication cables. Having this information on an overlay helped overcome the farmers' natural inclination to let existing fences and other man-made improvements restrict free thinking about possibilities for change.

PLAN SYNTHESIS

New layout

A second overlay (overlay B), was placed over the base plan and overlay A was flipped aside. Future developments based on the natural boundaries of soil type, slope, aspect, vegetation and drainage, bearing in mind the relevant enterprise(s), were sketched on overlay B.

We had to try not to be influenced by existing fencelines. This was more difficult for landholders whose families had been on the same farm for a long time than it was for more recent purchasers like Ross Kitchin and David McDonald or project staff. In most instances, input from John Marriott and me was required to start this process but, once started, farmers tended to plan with enthusiasm. For a couple of the demonstration farmers, refining their plans is now almost a form of relaxation. Although we attempted to ensure that each paddock on the new layout contained only one land type, this was not always possible, as we also had to consider what paddock sizes best suited the farm enterprise and any likely changes to that enterprise, and the management style of the landholder.

For example, on 'Reedy Creek' most paddocks were more than 40 hectares and many contained more than one land type when the plan was prepared, but Barry Levinson's enterprise and labour structure was based around these large paddock sizes. The process of delineating land units and comparing these with existing layout exposed instances where the existing layout was not ideal, but in Barry's case it was not possible to change all of these within three years. The advantage of the system is that Barry now has identified those fences which would ideally be located elsewhere, and will relocate them when they need repairs.

The plans locate fences along natural boundaries (particularly the critical ones of slope, drainage and soil type) where possible, so that each piece of land can be managed appropriately for both short-term and long-term production. The benefits of fencing to land type were exemplified on 'Satimer' where, in 1984, the large 'Sorrel' paddock was rectangular in shape, running north-west to south-east, with a high ridge down the centre. In summer, sheep tended to camp on the ridge, eating pastures to the ground, while pasture at the bottom of the slopes grew rank with undergrazing. With south-westerly winds in winter, one slope was much warmer than the other, but Bill and Jack Speirs had no control over which side of the ridge the sheep were on at any time. This was an obvious case for the top of the ridge to be fenced out and planted or seeded to indigenous trees, thus creating two paddocks, each with an even slope, one facing south-west and one facing north-east. Jack and Bill now have the ability to ensure more even grazing pressure, the top of the hill is no longer vulnerable to overstocking and wind erosion and sheep can now be held on the most sheltered slope when necessary.

Fencing to natural boundaries of slope, aspect, soil type, vegetation and drainage maximises the ability of the manager to achieve optimum grazing pressure on each management unit. The most productive land types can be more heavily stocked, and the more sensitive areas can be 'spelled' if the land is fenced appropriately, with a consequent increase in overall carrying capacity and a reduction in the potential for erosion due to grazing.

One drawback often cited as a disincentive to fencing to land type is that it often results in a greater number of smaller, odd-shaped paddocks. This obviously does not suit large properties extensively grazing big mobs with a minimum of labour or cropping extensive areas with large machines, using traditional management strategies. However an increase in the number of small paddocks can be used to advantage, as it was at 'Fernleigh' where a portion of the farm which had been subdivided into four paddocks was further subdivided into nine paddocks according to land type. The smaller paddocks, protected by shelter belts, suited Stuart Cuming's need for 'joining' and lambing paddocks for stud ewes in small mobs and greatly increased his flexibility of management. Regardless of the enterprise, smaller management units enable fine-tuning of management and more flexibility, which has to be weighed up against the cost of fencing.

In order to reduce land degradation and achieve sustainable production in extensively grazed areas, it is essential to move away from the traditional grazing methods towards flexible grazing systems based on smaller management units which reflect land types and which eliminate grazing pressure from those portions of the landscape where grazing is inappropriate. The Potter Farmland Plan demonstration farms have shown how this can be achieved, lifting production and avoiding land degradation.

This emphasis on fencing to land type is not just applicable to grazing enterprises in western Victoria. A farm layout which reflects land type also enables the most appropriate use of fertilisers, crop and pasture species, on either cropping or grazing farms. Different soil types require different fertilisers and respond to cultivation in different ways. It is very difficult to manage each land type according to its needs and capabilities when several land types occur in one paddock. Similarly, on irrigated farms, different soils have different infiltration and water-holding characteristics which should be reflected in the design of paddocks and irrigation bays and systems for the delivery and drainage of water.

One of the major features of the farm layouts on the demonstration farms is the development of laneway systems. The benefits of a well-planned laneway system can be illustrated using the following example from 'Gheringap' at Wando Vale. A mob of sheep in the far paddock is brought to the shed for crutching. Before implementation of the whole farm plan, Ross Kitchin had four paddocks between his woolshed and the far paddock, and a steep gully dissected by Corea Creek. In bringing sheep to the shed he had to first move the sheep in each paddock out of the way so that the mob to be crutched could be moved through without getting boxed (accidentally mixed together). He also had to take care with his dogs to ensure that the mob was in the right position to cross the creek over the bridge, rather than running along the edge of the creek and possibly attempting to jump down the very steep, eroding banks. On the way back from the shed to the paddock the process would be repeated, with the same degree of cursing, frustration and stress to sheep, dogs and man.

The laneway system, which now gives Ross access to each paddock from a lane, makes the above job dramatically faster and less stressful for Ross, the dog and the sheep. To get the sheep from the far paddock, Ross drives down the lane to the paddock, and the most work he and the dog have to do is to round up the sheep in the far paddock to get them into the lane, leaving the gate open behind him. Once in the lane, he simply follows them along, inspecting mobs in adjacent paddocks, rather than worrying about getting sheep boxed, with the dog resting on the back of the ute. At the creek, the sheep have no choice but to cross the bridge, as the creek on either side is fenced out and revegetated. Ross leaves the gate open from the shed to the yards after shedding the sheep for crutching. As he finishes each sheep, it can wander down the let-out race, into the lane and back to the paddock. By the time he is half-way through the mob, the first sheep is back in the paddock grazing, without any supervision or consequent stress. When Ross has finished the mob, he can have a cup of tea or do some small jobs before driving down to the paddock to shut the gate.

In Ross's case, the laneway more than halves the time and stress involved in moving sheep and eliminates the risk of boxing sheep or accidentally forcing them into the creek. But laneways have several other advantages. A three-row shelter belt has been planted on the southerly side of the exposed ridge which Ross's lane runs along. The shelter belt incorporates a row of Tagasaste (*Chamaecytisus palmensis*) which acts both as a bushy shrub and a 'living haystack', a source of fodder which can be periodically harvested when feed is short. The other two rows are indigenous eucalypts, wattles, sheoaks and paperbarks which provide shelter, habitat for native wildlife, and a potential source of timber in later years. Ross now has a sheltered strip perpendicular to prevailing cold winds, in which he can hold stock under cold, wet, windy conditions, particularly after shearing or during lambing. His laneway system forms the skeleton of a comprehensive network of shade and shelter, which doubles as wildlife habitat and a source of fodder and timber.

Ross also has the option of grazing the lanes very heavily in early summer to create a firebreak. With their adjacent rows of fire-resistant trees and shrubs, the lanes become very effective firebreaks. In the event of a fire, they can often slow down the fire sufficiently to enable tanker crews to work directly at the head of the fire. Another benefit of lanes during a fire is that they ensure much better access around the farm for crews unfamiliar with the farm layout, usually acting in conditions of haste, stress and smoke.

A critical factor which often dictated the extent to which it was possible to be innovative in creating new subdivisions on the demonstration farms was the design and expense of fencing. Fifteen farmers in a discussion about the 'best' fence, will usually have fifteen different opinions and this was the case on the Potter farms. Fencing is one of the traditional aspects of Australian agriculture in which old ideas die very hard indeed, and where the 'because we've always done it this way' philosophy is still healthy. We constructed fences largely to the farmers' preferences in the first year, but as we were able to demonstrate the effectiveness of modern designs we were able to modify some landholders' ideas in favour of more efficient fences (see Chapter 5). This enabled the implementation of land-type subdivision which would have been prohibitively expensive using traditional methods.

General Guidelines for Fence Location

In deciding where fences should be located within an overall philosophy of fencing to land type, we found the following guidelines useful:

- *Fence location:* Fences should either follow or be at right angles to the contour, as fences at an angle to the contour can trap and divert water, leading to erosion.
- *Corners:* We tried to avoid sharp angles so as to avoid erosion from stock moving around sharp corners and to ensure that machinery could work effectively. However in several instances, notably at 'Helm View', it was easier to retain a sharp angle, but to fence off the corner and plant it to trees, providing shelter, wildlife habitat and a potential source of farm timber.
- *Drainage lines:* Most drainage lines were considered to be a separate management unit, and most of the demonstration farm plans entail fencing and revegetation of drainage lines. This was done immediately at 'Helm View' and 'Reedy Creek', where the drainage lines were subject to erosion.
- *Ridges:* We attempted to locate fences as close to the top of the ridge as possible so as to separate different aspects, but at 'Satimer' and 'Reedy Creek', where the ridgeline was ill-defined, the high ground was considered to be a separate land unit. These sandy, erosion-prone hills were fenced out and planted to wood lots.
- *Laneways:* Laneways are subject to heavy traffic and are focal points for erosion. Where necessary on the demonstration farms, laneways were formed up into a crest with suitable drainage and water disposal points. Laneways on the contour, or in steep areas where they have to be sidecut across a slope, should be carefully surveyed and aligned so that run off can be channelled into an appropriate waterway and so that the laneway itself is least likely to erode.
- *Gates:* When planning farm layout it was essential to think carefully about the location of gates as sometimes the ideal location of a gate may alter fence alignment. Gates should be on high ground and stable soils, or reinforced with bedlogs and gravel, as they concentrate stock movement, and stock tracks near and through gates can be an erosion hazard. We

also considered the frequency of use, and type of machinery for each gate, particularly into tree areas.

After we had worked up a new layout with the farmer, the existing layout (overlay A) was flipped back on to the base plan to compare the new design (overlay B) with the old. Fencelines inappropriate to natural land units stood out. Where the difference between the 'ideal' and the existing layout was very small, the new layout was altered slightly to suit the old, recognising that the effort (in labour and money) involved in shifting fences or other features was not justified. However there is a case for retaining one overlay depicting the 'dream' layout, in case there is ever an opportunity to start again from scratch—for example, after a fire. Fences on both 'Nareeb Nareeb' and 'Fernleigh' were extensively destroyed in the 1977 fires and reconstruction after the fire offered opportunities for improved farm layout which were taken in both cases. Having a plan in the desk drawer outlining a well-considered 'ultimate' layout would be a great advantage in this situation, especially since, with dead stock to bury, boxed mobs of sheep and cattle to sort, fodder to procure and feed out and often teams of volunteer fencers arriving within days, there is very little time for long-range planning in the aftermath of a fire.

Water Supply and Drainage

One of the key features of the whole farm planning process used on the demonstration farms was that the ecological implications of all interactions were examined. Improvements such as trees, layout, pastures, salinity control and water supply were not considered in isolation. In going through the process on paper it is necessary to break it down into a series of steps, even though in practice, water supply was in our minds while drawing a new layout and discussing it with the farmers, as was access around the farm and sites for trees. So while this section considers the planning guidelines we used for water supply and the following section considers revegetation, it is important to realise that in practice, the distinctions between different aspects of the plan were blurred. This is inevitably the case for any farmer planning his or her own farm.

Water supply and drainage were carefully considered in creating the new layout, as they often influenced fence location. The farmers already had a good idea about whether their existing water supply was adequate for their present needs. We re-examined water supply in the light of the new layout to see whether it was compatible with the proposed changes and if it was capable of satisfying increased demand for water in the future and marked in improvements on overlay B.

The second of our whole farm planning principles was that each natural land unit should be managed according to its potential and its limitations. The land units which were identified immediately on every farm were the creeks, rivers and drainage lines. We knew at the beginning of the project that these would require close attention in the planning process. Natural watercourses are critical to the ecological stability of the landscape. In most agricultural areas, the quality of water in streams and rivers is declining and streamside environments are severely degraded. This degradation is due to overclearing and inappropriate management (such as overgrazing or the introduction of pests) in catchment areas, leading to erosion and subsequent high sediment loads in streams. It is also due to clearing of the riparian vegetation itself by grazing stream sides, leading to bank erosion and reducing the ability of the stream-side environment to filter out pollutants such as fertilisers and pesticides. Stock tracks to and from creeks and rivers become point sources for run-off which is often a starting point for gully erosion. Rising water tables lead to increasing salinisation in streams, which inhibits indigenous vegetation and restricts the potential for plants to stabilise stream beds against erosion.

The following examples illustrate how these issues were tackled on the Potter Farmland Plan farms in a way which benefited the farmer as well as the stream.

At 'Gheringap', Corea Creek is a reliable source of high quality water. 'Gheringap' is quite hilly, with tableland plateaus descending steeply into Corea Creek and its tributaries. When the Potter Farmland Plan started, the creek was the main source of water and stock had to move up and down steep slopes to drink. Gully erosion was active in several places on the creek bank. On Ross and Annabelle Kitchin's whole farm plan, the creek and its tributaries were identified as a separate land unit. There was an obvious need to fence out the creek and stabilise the banks with vegetation, but this involved creating an alternative water supply. We investigated constructing several dams higher up the slope. Ross was worried about the possibility of dams leaking, which is common in the local area due to the dispersibility of the clay subsoil.

We consulted Bill Sharp and Peter Dixon from the Department of Conservation, Forests and Lands, who were experienced in developing farm water supplies, about the feasibility of constructing a suitable dam for a reticulated water supply. A back hoe was hired to dig several test pits in likely dam locations. Peter Dixon sampled and analysed the subsoils and concluded that dam construction would be successful, provided that dams were well designed and carefully constructed in ideal soil moisture conditions, using a rubber-tyred scraper and a sheep's-foot roller, rather than a bulldozer working on its own in dry conditions. Quotes from the local contractors with suitable equipment were obtained. The dam option would have been expensive and difficult to organise in advance to anticipate and take advantage of optimum soil moisture conditions. Ross was still uneasy about dam construction, so we agreed to investigate other alternatives.

Peter Dixon suggested a second option, which was to use a hydraulic ram pump in the creek to pump to a high tank for subsequent reticulation to troughs on lower slopes. John Marriott, Ross Kitchin and Peter surveyed levels in the creek bed and found an ideal site for a hydraulic ram, where the creek dropped four metres in a horizontal distance of 50 metres. They then surveyed up the slope to the desired tank location (nearly 100 metres vertically above the creek) to ascertain the necessary 'head' required from the pump. Peter Dixon investigated suitable pumps and calculated the daily quantity of water required to ensure a safe supply over the number of sheep and acres in question, and the necessary tank size. He recommended the type and model of pump to buy and suggested a 30 000 litre concrete tank for a site near the top of the hill.

Ross constructed the site for the pump and installed it in the creek, assisted by Peter Dixon's plans. He ran an underground 'poly' pipe to the top of the hill and tested the flow from the pump for a few days. There were teething problems, as the flow from the pump was erratic. After consulting Peter Dixon and the pump manufacturer, Ross and John Marriott put a better strainer on the inlet pipe in the creek and an overflow tank between the 75-millimetre PVC collector pipe and the 500-millimetre galvanised delivery pipe to the pump. The pump seemed to function more evenly, but still stopped after several days. Ross then tested the flow at a point about 15 metres vertically lower down the hill and the pump worked beautifully. With subsequent fine tuning such as stabilising the delivery pipe and putting a lid on the small overflow tank feeding the ram, the system has reticulated high quality water for 24 hours a day, 365 days a year for more than four years with no energy input other than the flow of the stream and gravity, at a cost considerably less than that of several new dams. It has the added advantage of allowing Ross to locate troughs at places in paddocks where they are most convenient for inspection, for ensuring efficient grazing patterns and for minimising erosion.

This example illustrates the benefits of seeking expert advice and of Ross persevering to implement a sound principle in a practical way.

'Wyola' also relied on creek water for stock before the Potter Farmland Plan. Gavin Lewis was aware of some erosion of creek banks through stock traffic, and that the creek was becoming more salty. In the whole farm plan the creek was identified as a separate unit. When we tested the water, we found that it was about 5000 parts per million of salt at the end of summer, which is not critical, but marginal for young sheep and lactating ewes. It was obvious that in the long term an alternative was necessary. There was an additional incentive to fence along the creek in Gavin's case, as the paddock layout was such that several land types were in one paddock and grazing patterns were uneven.

The soil types on and around 'Wyola' are very reliable for dam construction, so we decided to construct a new dam, which Peter Dixon surveyed and designed in 1986. The 4500-cubic-metre dam was designed to supply water for four paddocks and was located as high up the slope as possible to avoid salinisation, which is common for dams located in gullies in the Melville Forest and Bulart districts. It incorporated a 50-millimetre PVC pipe through the bank, so that water could reticulate by gravity, avoiding the air pockets in the pipe which can sometimes develop where pipes run over the top of dam banks. The dam was fenced out and trees were planted around it. Sedimentation of dams and bogging of stock when water levels are low is eliminated when dams are fenced out and surrounded with a buffer zone of trees, shrubs and grasses. The trees around dams cut down on evaporative losses and are valuable refuges for native birds, insects and small mammals. As at 'Gheringap', erosion around watering points was minimised by careful location of troughs and surfacing with gravel.

Guidelines for Planning Farm Water Supply

In deciding whether the volume and quality of existing supplies, layout and disposal of water are adequate and in designing improvements, the following points were considered on the Potter Farmland Plan farms. They are useful guidelines for planning farm water supply.

- *Source:* There may be several, such as underground water, permanent streams or surface run off. If there is a choice of supply, the supply which fits in best with the new layout should be considered, as well as the cheapest or most reliable supply.
- *Quality:* If some dams are located in gullies which are becoming or have the potential to become saline, or if the system relies on natural water courses which are threatened by salt or other quality problems, then alternate supplies should be investigated, as discussed above. Water quality on grazing farms can often be improved by fencing off dams and creeks from stock and either reticulating or restricting access to suitable points.
- *Storage:* A few large, well-sited, fenced-off storages reticulating to many paddocks is often better than a small storage in each paddock. Whether you have several large storages or a smaller dam in each paddock, the

critical point is that they should be reliable. Depending on the reliability of rainfall, dams may have to be capable of serving the farm for several years without rain.

- *Management:* Absentee landholders need a system which minimises the need for inspection and maintenance. A dam in each paddock is likely to be more suitable than a reticulation system for absentee owners.
- *Watering Points:* In large paddocks it is desirable to have more than one watering point. Future trough sites should be marked on the plan even if troughs cannot be installed immediately. Watering points can be placed in such a way as to encourage stock to camp in less sensitive areas.
- *Waterways:* These can be used to control run off to prevent erosion and also to divert water into large dams to improve water storage. The capacity of the waterway should be based on the estimated (over 10 to 20 years) highest daily run-off from its catchment.
- *Erosion:* Waterways or drainage lines are a separate management unit and may need to be fenced on both sides and revegetated, as was done at 'Satimer', 'Gheringap', 'Warooka', 'Helm View', 'Willandra', 'Wyola', 'Daryn Rise', 'Fernleigh' and 'Reedy Creek'. If the watercourse is the sole water supply, it is important to prevent stock from damaging banks and muddying water. This can be done by fencing out the watercourse and pumping to tanks or dams higher up, or by restricting stock access to well-paved and fenced watering points.
- *Drainage:* Poorly drained areas should be examined to see whether they can be made more productive through better drainage as long as it is possible to dispose of the water without causing problems elsewhere. This is one of the future developments planned for 'Helm View'.
- *Wetlands:* If long-term production cannot be enhanced by draining poorly drained areas, then opportunities to drain excess water into them should be taken to maintain their integrity as wetlands for fire protection and wildlife habitat, as is planned for the large swamp at 'Satimer'.
- *Advice:* Expert local advice on farm water supply should be sought from experienced locals, soil conservation advisors or private consultants.

The Potter Farmland Plan demonstration farms are all dry-land grazing enterprises with only an occasional crop grown for stock feed. Consequently, the water-supply works carried out during the project were consistent with the demands of these enterprises rather than with those of irrigation farms. Farm plans for irrigated properties obviously place a much greater emphasis on water supply as the major constraint and the design of irrigated properties is influenced primarily by the movement and regulation of water.

However, on irrigated farms it is easy to fall into the trap of considering water supply and regulation to the exclusion of all other factors. Farm plans for irrigated farms should consider much more than just water and its delivery on to the farm and drainage away from it. In irrigated areas it is equally important to try to achieve an ecological balance, as hydrologic cycles have been altered so drastically through human regulation. Interaction between groups of adjoining landholders has an important role to play, particularly with drainage schemes and revegetation projects. Needs such as shade and shelter for stock and crops, wildlife habitat and landscape improvements should be considered before major works such as laser grading, as it is hard to incorporate these issues into the plan afterwards.

Land Management

After preparing a new layout, we re-examined the management units (paddocks) defined on the new layout. Possible roles within the farm enterprise were defined for each paddock according to the constraints identified. Some paddocks were specialised, others versatile, capable of and suitable for sustaining many land uses.

We found it valuable to break the plan into sketch maps of each paddock. This was done by tracing the layout from the plan using a black pen on plain white paper, which was then photocopied many times to keep the recording system up to date. For each of the Potter Farmland Plan demonstration farms, John Marriott kept folders full of these paddock plans detailing all operations, and inputs of labour and materials. It was a basis for costing and budgeting the implementation of the plan. For the average farmer, such a system is equally valuable for paddock records such as stocking rates and cropping and fertiliser histories.

One of the key long-term issues affecting land management in the Potter Farmland Plan demonstration areas is that of dry-land salinity, and the influence that the goal of salinity control had on the planning of the farms is an example of the consideration which needs to be given to land management over the whole farm, as well as to each land unit separately.

The background against which the Potter Farmland Plan approach to addressing dry-land salinity on the demonstration farms was developed is outlined below. This was an area in which we relied on our own knowledge and experience, advice from specialists in DCE and DARA, and research and extension activities in other regions and other states, to assist the Potter farmers to tackle salinity within the context of their whole farm plans.

The three demonstration areas are in nominated 'hotspots' for dry-land salinity within the state salinity control programme administered by the Victorian Salinity Bureau. Whilst little is known about the hydrogeology of the region in comparison with the irrigation areas and north-central Victoria, the causes of the problem are the same. Clearing of perennial trees and pastures and their replacement with shallow-rooted annual grasses has led to increased accessions of rainfall to the groundwater. As a result, water tables have gradually risen over many decades, bringing with them dissolved salts from former marine sediments and cyclic salt washed in through rainfall. When saline groundwater rises to within approximately two metres of the soil surface, water moves up through capillary action and evaporates, leaving salt behind which gradually accumulates in the topsoil. Salting first shows itself when clovers and other improved pasture species, such as ryegrass, are replaced in the pasture sward by salt-tolerant species in low-lying areas or at particular pressure points where groundwater emerges at the soil surface. The usual succession is Sea Barley Grass, followed by Buckshorn Plantain and Yellow Buttons, none of which are productive species. With prolonged salting, soil structure deteriorates, often developing an impermeable, mottled clay layer which impedes drainage and leads to waterlogging, which further restricts the ability of plants to tolerate salt in the soil. As stock tend to camp on these areas, they become bare very quickly and the water which flows naturally along these drainage lines is often the final agent in the erosion process.

Scientists from the Land Protection Division of DCE at Bendigo have been carrying out research into dry-land salinity at Neil Lawrance's property at Gatum, 10 kilometres north-east of 'Helm View', for many years. They are convinced that the process of dry-land salting, which is occurring in each of the demonstration areas, can be characterised as a 'Province A' process, which means that the groundwater in a given catchment or subcatchment is 'recharged' from within that catchment. Recharge areas are portions of the landscape in which water is soaking through the soil profile into the groundwater, as opposed to discharge areas, which are where groundwater is forced to the surface. In a 'Province A' process, the contours of the water table approximate those of the soil surface and it is theoretically possible to control the level of the water table by land-use changes within the catchment. This is a more favourable situation from an individual farm viewpoint than 'Province B' salinity processes, in which regional groundwater systems operate, where the water table may be recharged hundreds of kilometres from the subcatchments in which salting shows up at the surface. 'Province C' catchments are more complex, characterised by a mixture of Province A and Province B processes.

In the catchments of north-central Victoria, where most Victorian dry-land salinity research has been done, 'recharge areas' have been clearly defined. These areas tend to be highly fractured rocky ridges where rainfall

• It is much easier and less stressful for all concerned to move stock in laneways. Peter Waldron (below) is able to reach almost all his paddocks from lanes, which means that he is always only one gate away from home or the woolshed, thanks to the whole-farm plan. Ross Kitchin's laneway (right) gives him easy access to all paddocks on his steep, undulating farm. Mobs of sheep can be held in the lane, where they are protected by shelter belts. Laneways also greatly reduce the time spent mustering and the stress on people, stock and dogs, as Peter Waldron's lane shows.

• The 'sorrel' paddock on 'Satimer' was a large paddock with a long, central ridge in 1984. Sheep tended to camp on the ridge, leaving it bare and exposed. The whole-farm plan divides the paddock with a shelter belt along the ridge, creating two paddocks, each with a uniform aspect, one sheltered from the north and one from the south. Bill and Jack Speirs now have much better control of the grazing pressures in these two paddocks.

● When David and Russ McDonald bought 'Daryn-Rise', the gully above was just touching the boundary fence. Six years later, in 1985, it had moved 50 metres into the property and was still on the move. In May 1987 we installed a gully head control structure made of 'Black Brute' PVC donated by Hardie Iplex. Four men completed this job in one day and, along with the complementary new pastures, trees and more appropriate fencing higher in the 200-hectare catchment, it has completely healed a nasty wound.

• This shot dramatically illustrates the potential to rehabilitate saline, waterlogged areas with salt-tolerant species. Tall Wheat Grass has been successfully established on the left, preventing erosion and achieving some production from the saline area. The area on the right continues to degrade and is a liability.

• We were opportunistic in collecting seed from local provenances of healthy trees and shrubs. The author collecting *Bursaria spinosa* seed (at left) and (below) the Cavendish crew of the then Department of Conservation, Forests and Lands collecting *Acacia melanoxylon* seed.

• Planting rows 3 metres apart enables a tractor and slasher to work between the rows for fire protection in early years if necessary. Peter and Julie Waldron of 'Willandra' won a regional fire prevention competition in 1987, as their shelter belts were regarded as a fire-protection asset.

• This island of land, cut off from the rest of 'Cherrymount' by two gullies and the road, was ideal for a wood lot of eucalypts, acacias and casuarinas, which will provide a source of farm timber in the long term and wildlife habitat and landscape improvement immediately.

can infiltrate at rates of several metres per day, as opposed to lower slopes where infiltration rates are typically only several centimetres per day. In other words, a small proportion of the landscape contributes a high proportion of the recharge, which simplifies the options for control. The best example of such a catchment is at Burke's Flat near St Arnaud, which has been the subject of Soil Conservation Authority and now Land Protection Division research for many years. Water tables have been lowered more than four metres within five years under a regime which has seen the rocky ridges revegetated with trees, the upper slopes planted to phalaris and the lower slopes planted to dry-land lucerne. This regime has attempted to maximise plant water use, and is the on-ground success story of the Victorian salinity programme to date.

In south-western Victoria the recharge areas are not clearly defined and the few infiltration rates which have been measured by Peter Dixon suggest that infiltration is uniformly low across the landscape. This means that the accession of water to the water table cannot be easily simplified into the 'recharge/discharge' story. Recharge is potentially occurring across virtually all those areas that are not showing the effects of salt—every area which is not a discharge area is potentially a recharge area. Dry-land salinity control in the demonstration farm areas must therefore attempt to maximise plant water use over the entire landscape, using every drop of rain where it falls, or ensuring that it runs off the farm safely, rather than soaking into the water table. Under the present grazing enterprises, a strategy should in-corporate vigorous perennial pastures, it should ensure an even distribution of water to those pastures, it should facilitate safe run off of water not used by plants and the pastures should be complemented by deeper rooted, perennial vegetation such as trees and shrubs.

Plant water use had little influence on the location of trees on the de-monstration farms, although common sense suggested that plantations on ridges and 'along the break of slope' were useful starting points in terms of the effectiveness of using water through transpiration. Plantations on the margins of saline areas, sited to shelter small paddocks of salt-tolerant pas-tures, have the potential to create a cone of depression in the water table immediately beneath the trees, and thus assist in restricting the expansion of the salt-affected area. In the main, however, trees were established where they were of most benefit to the farm operation.

Perennial pastures such as Phalaris and Perennial Ryegrass were incor-porated in whole farm plans and established wherever possible and trees were established where they could be of most value to the farm, in terms of shade and shelter, wildlife habitat, fire protection, landscape improvement and timber production.

For example, on 'Daryn Rise' deep-rooted perennial pastures, which directly increase farm production, including Phalaris, were sown on the higher slopes and ridges, salt-tolerant pasture species (including Tall Wheat grass, Palestine Strawberry Clover and Puccinellia) were sown on saline areas and a network of predominantly indigenous trees and shrubs was established across the farm. The trees complement the pastures in using water, while providing other benefits such as shade and shelter for stock, farm timbers, wildlife habitat and landscape improvement. Future plans include surrounding salt-affected areas with trees as was done at 'Helm View' and 'Willandra' at Melville Forest. Changes to farm layout to reflect land types enabled better establishment and use of pasture, facilitated treatment of erosion problems and improved access around the farm. Provision of new water supplies minimised erosion of natural watercourses and ensured better quality water for stock. Establishment techniques for trees and pastures are discussed in more detail in Chapter 5.

REVEGETATION

The farmers were very aware of the need for more trees and at the start of the project many regarded trees as the most important farm improvement associated with the development of whole farm plans. As the planning process developed, the appropriate place for trees in the sequence of plan preparation became apparent for these farmers.

We did not outline revegetation on the whole farm plan until a new layout and water supply improvements had been designed although, as mentioned above, the ecological need for revegetation and the productive benefits of well-planned farm trees were in mind all the time as new layouts were designed and discussed. The guidelines vary for different types of farms; for example, a vegetable enterprise has a different requirement for trees and shrubs from a wheat farm, but there are opportunities for increased tree and shrub cover on all Australian farms and in order to increase the ecological diversity in agricultural areas, it is essential that these opportunities are taken up as soon as possible.

'Trees' are used in the broadest sense in the Potter Farmland Plan, referring not only to tall species, but including communities of indigenous trees, shrubs, herbs and grasses. Species are regarded in groups rather than separately, as in many cases species which naturally form part of a community of trees and shrubs will live longer, will provide better habitat, and will be more resistant to pests if they are planted in the same groups, rather than alone or as trees without understorey companion shrubs.

Tree and shrub communities were outlined with the new layout and water supply on overlay B. The steps for incorporating trees into the whole farm plan which we followed on the demonstration farms are described below.

Role of Trees and Shrubs

Our first step in planning tree establishment was an examination of the function the trees were expected to fulfil. No realistic assessment of species, site preparation, establishment technique or layout can be made without a clear appreciation of what trees are to do. This illustrates the need for a whole plan to be prepared which places any revegetation in context and makes the role of trees very clear. The various roles of trees outlined below were not considered as a series of separate functions, but as overlapping parts of a whole—as revegetation of a farm must be looked at in a holistic way in order to maximise benefits to both the landholder and the land.

Protecting and enhancing remnants

The whole farm plans outline the protection of remnant indigenous vegetation from grazing or other disturbances, where feasible within the farm enterprise. Several of the plans also specify works aimed at maintaining remnant stands in a healthy state through encouragement of regeneration. These remnants are a valuable genetic resource, and are extremely important to wildlife habitat. An example of treatment of remnant vegetation is at 'Daryn Rise', where the only native trees remaining on the farm in 1985 were in a small scraggly clump of *Eucalyptus obliqua* on a sandy knob. David McDonald fenced these trees off in 1986 and slashed the bracken underneath them. If there is no regeneration, the area will be burnt and some local seed from understorey shrubs will be scattered over the site.

Natural watercourses

Stream sides on the Potter farms were fenced and protected with indigenous vegetation wherever possible, to minimise erosion, improve water quality, and maintain the integrity of creeks and rivers, which are genetic corridors of extraordinary riverine and terrestrial diversity, vital to the health of the whole landscape. We fenced out dams and creeks on 'Wyola', 'Reedy Creek', 'Warooka' and 'Helm View' and surrounded them with trees and shrubs to improve water quality, reduce evaporation and siltation and provide wildlife habitat. We took care with species selection and with planting design to avoid the risk of tree roots breaching dam banks, to leave flight paths for birds and to avoid sheltering windmills.

Wetlands such as swamps are very important elements in farmland ecosystems. Opportunities to preserve the integrity of such areas by restricting stock access and re-establishing indigenous vegetation should be taken. At

'Satimer', the whole farm plan outlined the fencing out and revegetation of a 10–hectare wetland, however this part of the plan was not implemented during the first three years of the project. Bill Speirs is trying to obtain government and corporate assistance to improve the habitat value of the Satimer swamp with revegetation, nesting boxes and floating islands to complement the demonstration value of the rest of the Potter Farmland Plan works programme.

Erosion-affected areas

On 'Daryn Rise', one gully was actively eroding and the head of the gully had broken through the boundary fence and moved nearly 50 metres into the property in between 1979 and 1985. We began by fencing out the entire drainage line and asking Peter Dixon for assistance in designing a trickle pipe structure to control erosion at the head of the gully. Peter designed a structure involving two earthen banks diverting water into a 1.2-metre wide corrugated PVC tank, from which a corrugated PVC pipe (600 milli-metres wide and 18 metres long) directed the water into the bottom of the gully. The PVC material, called 'Black Brute', was donated to the project by Hardie Iplex. It is exceptionally light and strong and does not rust, which makes it ideal for erosion control works in saline drainage lines. The cor-rugated external surfaces and smooth internal surfaces minimise the potential for water to leak along the outside of the pipe causing further erosion. Five of us assembled and installed the structure in a day. The edges of the gully were battered and the wing walls were constructed by Doug Heard later. The area was planted and seeded to salt-tolerant species of trees and pastures (refer to the photographs in the colour sections).

In other instances at 'Satimer' and 'Reedy Creek' erosion was addressed by improved farm layout. Trees were used for erosion control in areas too steep or too severely eroded for grazing at 'Satimer', 'Gheringap', 'Helm View', 'Will-andra', 'Warooka', 'Wyola', 'Reedy Creek', 'Daryn Rise' and 'Cherrymount'.

Erosion control pays. The soil lost in wind and sheet erosion usually has the highest nutrient content and other problems such as gully and tunnel erosion not only remove valuable topsoil, but create management problems, restricting access and influencing layout. Australian soil is effectively non-renewable and we cannot afford to keep losing it. Trees can protect topsoil and minimise soil loss. However careful planning is critical, as poorly located trees can concentrate stock movement to contribute to erosion problems and poorly designed shelter belts in sandy soils can also increase wind-speed through gaps, especially in gateways, and exacerbate wind erosion.

Salt-affected areas

Trees were only one element of the Potter Farmland Plan salinity-control programme. The main benefit of trees for salinity control is their ability to transpire water from deep in the soil profile all year round, which gives them the potential to lower the water table. Trees established on sandy slopes and ridges on 'Reedy Creek', 'Daryn Rise', 'Helm View', 'Satimer', 'Gheringap', 'Fernleigh' and 'Cherrymount' are likely to reduce water penetration in their respective catchments and the trees planted and seeded on the margins of saline areas at 'Helm View' and 'Willandra' will also lower the water table in their immediate vicinity, which will potentially limit the lateral expansion of saline discharge areas.

Wildlife habitat

Much of the wildlife that occurs on or can be attracted to farmland is beneficial to the farmer and the community. Opportunities should be taken, where possible, to link areas of vegetation to provide better habitat for native wildlife, particularly birds and insects, which can help to control pasture and crop pests. Well-designed shelter belts of indigenous species on all of the demonstration farms are potential wildlife corridors, as are erosion-control plantings and stream-side vegetation. This was dramatically demonstrated at 'Satimer' when the 1986 revegetation assessment crew placed six small mammal traps in the shelter belt which had been established for only eighteen months. Next morning five of the traps were occupied, four by native marsupial rats and one by a blue-tongue lizard. It is obvious that colonisation of these areas is very rapid. Farm wood lots or occasional lambing havens such as those at 'Reedy Creek', 'Helm View', 'Fernleigh', 'Cherrymount', 'Willandra', 'Ballantrae' and 'Barnaby' are valuable habitat areas as they incorporate a balanced community of local trees, shrubs and grasses.

The most commonly cited benefit of wildlife on farms is that birds prey on pasture-eating grubs. But there are many other interactions between the wide range of organisms which should be considered under the term 'wildlife' and their environment. These interactions form a complex web and the balance between the various elements of the ecosystem has a great influence on the productivity of rural landscapes. We are only familiar with a very small part of the total ecological picture in our farm environments, which in the main have been dramatically altered from their natural state, yet with each new piece of information the importance of conserving remnant habitat areas and re-establishing wildlife habitat on farms is underlined.

As well as birds and insectivorous native marsupials which prey on pasture pests, other insects help to keep the populations of potential pests in balance.

Insects, invertebrates and micro-organisms such as bacteria, fungi and algae all play important roles in nutrient cycling and formation of soil structure within native ecosystems. The genetic diversity of natural ecosystems confers upon them stability and the ability to recover from disturbance and stress.

On most modern farms, however, there has been a continual reduction in genetic diversity, caused by clearing native plant communities and replacing them with large areas of crops and pastures which consist of only one or a few species. The management practices of modern western agriculture, such as cultivation and the use of fertilisers and pesticides, have also reduced genetic diversity above and below ground and this adversely effects soil structure and nutrient cycling. Monocultural systems tend to be more vulnerable to pests and diseases and their ability to recover from such stresses without the external aid of pesticides and fertilisers is reduced.

In the long term, it is important to reduce dependency on external inputs of energy if agricultural land is to be managed in a more sustainable way. Natural ecosystems provide the clues about how we can introduce some of the checks and balances which stabilise natural systems into the design and management of farms. There is great need for research in this area, but while waiting for research, common sense suggests that it is essential that we must conserve those relatively undisturbed areas of native vegetation that remain in agricultural areas.

Where possible, native vegetation was re-established on the demonstration farms in a way which imitated the local species communities. In most cases, through careful design, we were able to do this so that revegetation works enhanced farm management and production by providing shade and shelter, fodder, wood products and so on.

Shade and shelter

It was clear that the primary value the farmers associated with trees was the provision of shade and shelter, with good reason. The Potter farms are in a predominantly grazing area, one of the richest wool-growing regions in the world. The climate is excellent for wool production, with reliable rainfall ensuring the development of productive perennial pastures on mildly acid but reasonably fertile soils. However the region has been extensively cleared of trees and shrubs during the last 150 years and many areas are now exposed and windswept. Cold 'snaps' with driving winds and rain, low temperatures and very high chill factors can occur at any time from May to January and shorn sheep and new-born lambs are very vulnerable during such conditions.

Well-planned farm shelter is not just a life-insurance policy for sheep in extreme conditions. Animals have a climatic 'comfort zone', within which they convert the food they eat into meat and fibre most efficiently. When temperature, wind speed and rainfall combine to create unfavourable conditions outside the comfort zone—either too hot or too cold—the animal uses energy in keeping warm or cool, which reduces the efficiency of food use, and consequently of production. Shelter belts have been established on all the demonstration farms. These shelter belts will eventually provide both insurance from extreme conditions and modification of the climate, effectively expanding the comfort zone of sheep and cattle and ensuring that feed is used more efficiently in cold conditions, particularly in areas where shelter for stock is critical, such as around sheds and yards, and lambing, off-shears or calving paddocks. Small clumps and/or single trees have also been sited on all the Potter farms to provide shade for stock in hot weather.

As well as perimeter shelter belts, clumps, shelter belts and single trees were also located *within* paddocks where they would be most useful. For example, at 'Reedy Creek' boomerang-shaped, mid-paddock shelter belts were planted in the north-eastern corner of one large paddock, where stock tend to be forced in extreme conditions and lambing ewes used them within eighteen months of tree establishment. Mid-paddock shelter designs are discussed in more detail in Chapter 5. Orientation, location, shape and structure are critical, both for protection of stock and for gaining all the benefits trees can provide. For example, trees must be arranged in such a way as to avoid creating wind tunnels or frost pockets and shelter belts can also be used to provide timber if they contain several rows of timber-producing species, rather than one or two rows of species which provide shelter only.

The creation of effective networks of farm shelter was a major feature of the Potter Farmland Plan operations between 1985 and 1987. 'Daryn Rise', a hitherto windswept farm, is a good example of the principles and practice of shelter, incorporating laneway and subdivisional shelter belts, wood lots, clumps and individual trees, taking into account the need for different types of shelter in different parts of the farm; and demonstrating how shelter can be integrated with other needs on farmland.

On the Potter farms, wood lots of up to eight hectares have been established, some of *Pinus radiata* and others of indigenous eucalypts, acacias and casuarinas, on a range of sites. At 'Reedy Creek', fuel-wood and pine wood lots were established on sandy hilltops which had been prone to wind erosion before 1985 and which 'blew' (suffered severe wind erosion) very badly in the drought of 1982-3. These very well-drained areas provide excellent emergency shelter. At twelve of the fifteen farms gully plantings and small wood lots complement shelter belts and clumps in providing emergency shelter.

Experience on the demonstration farms and the impact of a storm north of Hamilton on 1 December 1987, described in Chapter 5, led me to realise that there are two distinctly different types of shelter required on grazing farms—the shelter required to improve the day-to-day living environment for stock, and the shelter required to prevent stock deaths in extreme conditions—and that these different kinds of shelter must be considered as separate components of a farm plan. Through observation in the field and discussions with landholders, the following rule of thumb emerged: on grazing properties in southern Australia, it is a valuable insurance policy to have at least one 2- to 5-hectare area of trees, shrubs and grasses for every 2000 dry sheep equivalents, into which stock can be driven to gain complete shelter in extreme conditions. These areas can pay for themselves many times in one night by preventing stock losses. If no such areas become obvious when designing a new farm layout, then they should be deliberately incorporated. They should be placed close to where the most vulnerable stock are likely to be, preferably adjacent to a laneway for good access and on a well-drained site sheltered from the most likely direction of freezing winds. If all of these criteria cannot be satisfied, the priorities are access and drainage, since trees and shrubs can provide shelter regardless of aspect if the emergency shelter area is big enough.

Single trees, which occupy less space than clumps, were also planted on the demonstration farms, mainly in smaller management units where shade and shelter were required such as holding, lambing and off-shears paddocks. They were used around the ends of mid-paddock shelter belts too, at 'Daryn Rise,' 'Cherrymount', 'Fernleigh', 'Nareeb Nareeb', 'Ballantrae', 'Reedy Creek', 'Willandra', 'Pine Grove, 'Wyola', 'Barnaby' and 'Satimer' to reduce turbulence and to 'soften' straight lines, creating a more attractive farm landscape.

Single trees can also be used in avenue plantings as it is preferable for fire protection to keep a well-grazed or cultivated area around the homestead, particularly at points of entry and exit. We avoided having unbroken lines of fuel, such as shelter belts, leading to assets like the homestead (see fire section below), although there is a very good case for locating strategic shelter belts to the north and west of home paddocks to reduce wind speed and slow down fires, making direct attack on the fire more feasible for tanker crews.

Single trees and clumps were also intended to cause stock camps to disperse, in order to avoid the common problem of uneven grazing pressure and soil fertility resulting from concentrated stock pressure in favoured areas, which are often the higher areas most prone to wind erosion.

Fodder

There was no need for supplementary fodder on the demonstration farms where perennial pastures can easily be grown and where we have yet to learn how to make the best use of the grass we can grow. Fodder trees such as Tagasaste were incorporated into shelter belts at 'Gheringap' and 'Daryn Rise'. In other areas, such as the wheat-sheep belt of south-west Western Australia, trees can be a major fodder crop. Many landholders in Western Australia are now finding that they can carry more stock throughout the year because of the autumn feed provided by Tagasaste or Saltbush on formerly unproductive, sandy ridges. Fodder trees, nut trees and trees for cut flowers such as Banksias may also be incorporated into shelter belts, providing occasional feed supplements or a long-term cash crop.

Wood products

We decided at an early stage that the establishment of trees for commercial purposes on the demonstration farms would be done in such a way that wood lots complemented other aspects of farm operations, such as shelter, landscape and wildlife habitat. Potentially commercial wood lots were only established where the farmer showed a particular interest in learning to manage trees to produce a product, in the same way that pastures and stock are managed. The two most specialised wood lots established were of *Pinus radiata* and mixed eucalypt species suitable for firewood at 'Reedy Creek' and of *Eucalyptus camaldulensis, Eucalyptus saligna, Eucalyptus sideroxylon, Pinus radiata,* and Douglas Fir (*Pseudotsuga menziesii*) in an agroforestry design at 'Satimer'. These wood lots were very small, of more value for demonstration purposes than for a significant diversification for the farmers. In the main, the potential for timber production on the Potter farms will come from plantations established for multiple purposes, and this was our original intention.

Species with a commercial value were incorporated into shelter belts, wood lots and erosion control plantings on most of the farms. These species were mainly natives with a potential for use on-farm and for fuel wood, such as Red Gum (*Eucalyptus camaldulensis*), Red Ironbark (*Eucalyptus sideroxylon*), Spotted Gum (*Eucalyptus maculata*), Sugar Gum (*Eucalyptus cladocalyx*) and Swamp Yate (*Eucalyptus occidentalis*). With careful planning and some fore-sight, a farm-timber supply and an extra source of income can be in-corporated into the whole farm plan without removing much land from short-term production. Small wood lots can double as emergency shelter blocks and 'hard to manage' or unproductive areas may be revegetated with wood lots, agroforestry or clumps of trees and shrubs to provide wood, fodder, shelter or other products such as wildflowers, herbs, honey or essential oils, or simply to re-establish indigenous vegetation and improve landscape and habitat.

It is important to consider farm timber from a different point of view from that of a commercial forestry operation. Species which may not be economic for large-scale forestry companies may be ideal for a farmer who can integrate them into an existing enterprise and provide the necessary labour when convenient. It is also important to realise that the rules which apply today are unlikely to apply in ten to eighty years' time, when the trees are yielding products. Demand, markets and harvesting and processing equipment will also change. The degree to which trees are used commercially will depend on the enthusiasm and management ability of the landholder, the type of product and its market value. Local foresters and timber merchants can give advice on the latter two factors. The main point is that financial returns from timber are possible within a traditional farming system, but that they are additional to many other benefits if the trees are well planned.

Fire

Sensible design and management of trees on the demonstration farms has brought into question the belief of many firefighting authorities that native trees on farms are a fire hazard. On 'Willandra', more than 20 000 trees were established on less than 300 hectares in three years, yet the farm won a regional fire prevention competition, and judges were satisfied that the revegetation works were a valuable part of the protection strategy for the property.

On 'Willandra', as on the other demonstration farms, protection from fire was considered in the arrangement of trees and the selection of species, particularly for those near houses and sheds. Shelter belts of species that are resistant to fire and able to recover from it were located where they would protect assets by slowing down fire, making control more feasible. Within nearly all shelter belts, we ensured that the tree rows were sufficiently far apart to enable a tractor to work between the rows during the first two or three years, either slashing or ploughing long grass to create an effective fire break, and to protect the trees. Where possible we avoided having lines of flammable material such as trees and long grass leading to buildings from the main fire hazard. We tried to ensure safe entry and exit from the farms in a fire by avoiding avenues of flammable trees along driveways, although some of the farms already had driveway plantings of cypress and pine, which are potential fire hazards.

Landscape

The visual appeal of the farm landscape is very important and likely to become more so as more people seek a country lifestyle. The primary purpose of the individual trees and clumps of trees and shrubs used in the Potter Farmland Plan farms was to provide shade or reduce wind speed, but they also serve to soften straight lines. A landscape is more likely to be visually pleasing when there is a balance between remnant vegetation (particularly along drainage lines and ridges), shelter belts, clumps and individual trees of indigenous species. Since the basic framework of the whole farm plan reflects natural land types, the location of trees tends to emphasise elements of the landscape, borrowing from natural lines and spaces formed by boundaries of slope, remnant vegetation and water.

Living environment

A major difference between town and farm is that people don't just work on a farm, they also live there. The importance of an attractive environment in which to live and work is often underestimated on farmland, but it is increasingly evident in the premium prices being paid for visually attractive farms. Many of the participants in the Potter Farmland Plan mentioned it as a major benefit of the project. If planning is done according to the basic landscape principles outlined above, the result will be an *attractive* and *productive* farm landscape.

Although this discussion of the productive roles of vegetation has been divided into discrete sections according to the various benefits which balanced communities of trees and shrubs can provide on farms, it is important to emphasise that in a well-planned farm layout, trees are rarely planted or seeded for a single purpose. Trees on the demonstration farms are incorporated into the farm plan so that each planting provides a number of benefits, complementing other plantings, and also other farm improvements such as water supply, access and layout changes. In other words, trees should not be considered in isolation, nor should a tree programme be divided into different segments, each with a different purpose. With a whole farm approach, trees and shrubs can increase long-term farm productivity in a number of ways.

Species Selection

The need to maintain or increase biological diversity on farmland led project staff, encouraged by Bill Middleton, to select indigenous species wherever they were able to fulfil the desired functions. This was achieved quite simply, as participating landholders were involved in collecting seed from indigenous plant communities on or near their properties. This seed was either grown into tubed seedings by the Vicflora nursery at Wail, or used for direct seeding. Indigenous species are well adapted to local conditions (soils, climate and fire), are most likely to regenerate naturally, provide the best habitat for local mammals, birds and insects and help to maintain local genotypes, retaining the local character of an area.

There are some instances on the demonstration farms, as on many farms, in which the local environment had changed. In these cases the best species for the job was used, as long as it was suitable for the site. For example, we established non-local and non-indigenous species of trees (such as Swamp Yate) and grasses (for example, tall wheat grass) on saline areas; in wood lots (Red Ironbark, Spotted Gum); in fodder and shelter blocks (Tagasaste); and for erosion control (River Sheoak). In each of these instances local species were unsuitable, however the main species used on the demonstration farms were indigenous.

Site factors which influenced species selection included soil type and drainage conditions, seasonal weather conditions (not just how much rain but when it fell, and how many frosts of what severity) and elevation. These factors, combined with suitability for the desired purpose and the general rules outlined, were all taken into account in determining an appropriate species mix for the demonstration farms.

At first, John Marriott and I selected species with advice from Bill Middleton, but when the results of the first year of planting had become obvious, the landholders began to take more part in species selection. Several farmers expressed a dislike for, and resistance to, the use of Black Wattle, *Acacia mearnsii*, because it tends to be short-lived, untidy and very difficult to saw when dry; nevertheless Black Wattle and other indigenous acacias were included because they provide essential habitat and are useful in nutrient cycling. After four years the farmers were to have the last laugh because many of the Black Wattles, which had been growing very well, began to be severely attacked by the Wattle Fire Blight Beetle, *Poropsis*. Nevertheless the high organic turnover and nitrogen-fixing capacity of these fine-leaved bipinnate acacias may have served their pioneering role by this time.

After selecting species, we collected seed from a number of trees, preferably on similar soils, elevation and aspect to the proposed site. Nurseries need seed and seedling orders for species which cannot be collected locally, about a year in advance of planting, so it is important to think ahead in farm-tree planning. When we ordered seedlings which were not grown from local seed, we tried to ensure that they came from areas as similar to the planting site as possible.

Establishment Technique

One of the major features of the demonstration farms is that a range of establishment techniques was used on each farm, according to the requirements of specific sites (see Chapter 4). The main point to note when planning is to consider establishment techniques sufficiently far in advance of planting or seeding to ensure adequate site preparation. The basic techniques were planting (machine or hand), direct seeding (machine or hand) and natural regeneration.

Site Preparation

The way in which sites were prepared for trees on the Potter Farmland Plan farms is discussed in detail in Chapter 4, but it is important at this stage to note the time scale over which it was necessary to plan aspects of tree establishment such as species selection and site preparation. Timing is a major factor when preparing and implementing a whole farm plan. After two years of a three-year implementation on the demonstration farms, participating landholders had incorporated the sequence of operations required for their larger-than-usual revegetation programme into the rest of their farm operations. With a plan it was easier to see that seedlings should be ordered six to nine months before planting, that ripping should be carried out six months before planting, that herbicides needed to be sprayed one or two months before planting and that protection should be provided immediately after planting. Having a specified annual works programme was a big advantage and helped to eliminate bottlenecks in workload and poor sequencing of operations.

Protection

More than 80 per cent of the revegetation works carried out in 1985 and 1986 achieved a survival rate of over 90 per cent. The majority of failures were due to inadequate protection from stock.

Newly established trees are vulnerable to stock and other browsing animals, to insect and bird attack and to dessication. Good species selection, establishment techniques and site preparation can all be rendered irrelevant if the young plant is not protected. Planning for protection is critical from a financial perspective too, as we found that fencing and tree guards were easily the most expensive aspect of tree growing, as the diagram below reveals: Although the diagram overleaf refers to the entire programme's works budget, which was allocated to more than tree establishment, it is a reasonable approximation for the average farmer, who tends not to cost his/her own labour.

The special requirements of tree protection systems are discussed in Chapter 4. Like many farmers, most of the Potter Farmland Plan farmers

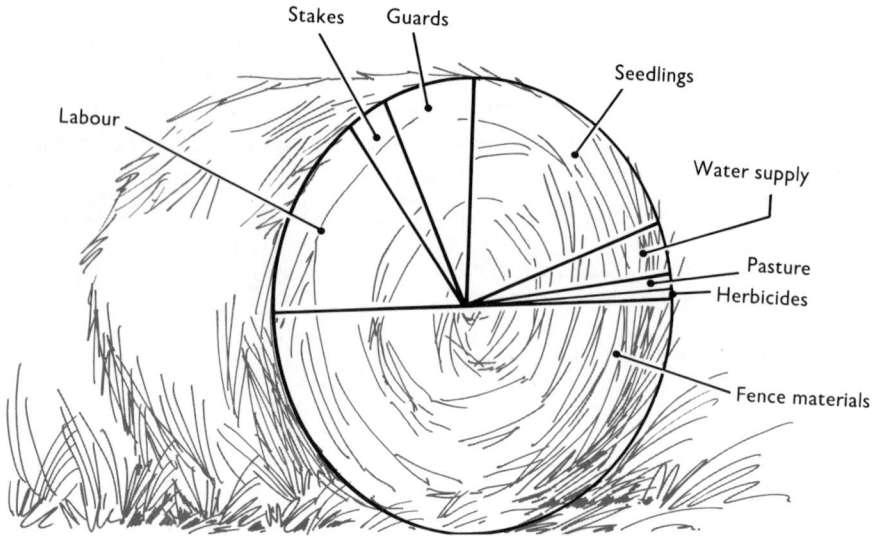

• Breakdown of Potter works expenditure, 1986

in 1985 did not recognise that fences to protect trees have to be better than normal subdivision fences. This factor influences the expense of implementing a revegetation programme and, consequently, the amount of a plan which can be implemented in any one year. Tree protection fences are often placed under more pressure than other fences. When an area such as a shelter belt has been fenced from stock for a season it usually contains more feed than the paddocks on either side, especially when feed is scarce, such as in the autumn, which is also when fence wires tend to be looser because they expand in hot weather.

An existing fence may be adequate for keeping different mobs apart when each paddock has a similar growth of pasture, but it is often inadequate for protecting young trees when there is a lot of grass inside the shelter belt and very little in adjacent paddocks. One of the most common situations in which trees are damaged by stock on farms is when a new fence is erected beside an existing fence to create a shelter belt, and the extra pressure suddenly placed on the older fence causes it to fail, resulting in tree losses. We had this problem at 'Barnaby' and 'Nareeb Nareeb', where old fences failed, rendering well-designed, new fences superfluous. Slack 'cocky's gates' or mallee gates across shelter belt ends are a false economy, as Barry Levinson, David McDonald and Peter Waldron found.

The decision to place a shelter belt along an existing fenceline is valid only if that fence is in the optimum position to provide effective shelter and effectively excludes animals. In other words, the investment of land, fencing and labour which goes into a shelter belt demands that it should be in the most effective position, protected on its full perimeter by fences of a high standard. If an existing fence is well located, it might be a matter of upgrading its quality to make it suitable for tree protection. This can be done most easily by adding one or two electric wires on 'outrigger' supports. Jeremy Lewis applied this technique successfully to reinforce an old fence on 'Barnaby' after it had failed.

These factors were all taken into account when planning and costing operations on the Potter farms to ensure that the proportion of the plan which was implemented each year was physically and financially manageable.

Maintenance

The need for maintenance of trees was almost completely avoided on the demonstration farms, through good species selection, planting design and site preparation, but it does have to be considered in the planning process. The maintenance of newly established trees can be tedious, time consuming and expensive, particularly if site preparation has not been thorough and effective. In a shelter belt, for example, if long grass is likely to present a fire hazard during the first summer it may be necessary to be able to work a tractor and slasher (or plough) between rows, which influences row spacing and means that a formal planting arrangement is required.

We did not need to water trees during the first summer, but this is a consideration which can also limit the number which can be established in any one year. Once again, good site preparation and establishment techniques can reduce the need for maintenance, but maintenance (and management practices such as thinning and pruning) must be included in costing and budgeting, as it is an integral part of any tree-establishment programme.

EVALUATION

After we had prepared the first draft of the whole farm plan with the farmers, we needed to assess the capacity of the farm and the landholder to implement the plan to determine how long it would take to put the plan into action and what resources would be required. Our experience on the demonstration farms proved that this process quickly exposed any inappropriate sections of the plan. It was an essential step to take before preparing any subsequent drafts. The sequence of costing, setting priorities and preparing a budget and works schedule outlined here was not quite followed on the demonstration farms. We deliberately set out to achieve in three years a programme of works which would take individual farmers at least ten years without assistance, in order to clearly demonstrate the whole farm planning approach over complete farms. This section outlines the process we used, while pointing out where a property acting independently would vary from this.

Costing

The first step in evaluating the first draft plan was to cost all proposed works accurately. The costing process should involve an assessment of likely benefits from each element of the plan so that priorities can be determined and the resources allocated. In costing works on the farms, the benefits of the elements of the plans were considered in terms of their value as demonstrations, as well as their benefits in productivity.

Assessing possible future benefits was very difficult compared with costs, which are usually more tangible and easier to measure. However it was worthwhile to attempt to predict and define benefits. The exercise emphasised how the different elements of the plan were interconnected and how important it was to consider the farm as a whole, rather than as an *ad hoc* collection of parcels of land. On the demonstration farms, this process became easier after one or two years' implementation. It was essential for the Potter Farmland Plan farmers, but it is also important for the average farmer to cost labour and time as well as materials, in order to get a realistic idea of the plan's impact on the overall operation of the farm.

PRIORITIES AND WORKS SCHEDULE

Overlay B showed the proposed improvements of the whole farm plan. It was useful to question the significance of each of these improvements and to rank them in order of priority.

To enhance the demonstration value of the farms, we established elements of each plan so that all of the priority areas listed below were represented, in order to show how each type of activity can best be integrated with the rest of the farm. As a general rule, the initial works programme concentrated on farm improvements which did not generate an immediate cash flow such as revegetation and changes to farm layout to reflect land types and protect sensitive areas of the landscape. The Potter Foundation funds were used to put the long-term framework of each plan into place with the understanding that the project participants would implement short-term pro-

• Before and after: the visual changes to this landscape have been subtle, but the fencing out and revegetation of the main dam at 'Helm View', the establishment of more salt-tolerant pastures and the replacement of dead and dying older generations of trees with vigorous indigenous communities of trees and shrubs will have a major long-term impact on the productivity of the land.

• Before and after: this sequence shows the visual impact of land-type fencing and revegetation on 'Helm View'. The later photograph was taken from the centre of a recently established mid-paddock clump of trees. Some decline in the few older living trees in the middle ground is also evident.

duction measures, such as pasture improvements, without assistance after the first three years of the project.

The operations of 1985 therefore emphasised revegetation and farm layout improvements. However, it soon became evident that this emphasis was causing the Potter Farmland Plan to be perceived as a 'large tree-planting exercise'. Consequently the emphasis changed slightly to demonstrate all aspects of whole farm planning, including measures with short-term production benefits. As a result, pasture, water supply and farm access improvements represented a greater proportion of Potter Farmland Plan operations in 1986 and 1987.

Suggested priorities for most farmers are as follows:

1 *Improvements which increase farm productivity immediately.* Pasture improvements, laneways, new crop rotations and changes in fencing are examples of measures which improve access or management and generate a cash flow to pay for implementing more of the plan.
2 *Measures which will immediately prevent further loss of productivity.* Examples of these are treatment of urgent erosion, salinity and tree-decline problems.
3 *Measures which gradually affect protection of land and other assets.* Increasing plant water use in recharge areas, drainage and water supply works, provision of shade, shelter and wildlife habitat are examples of this kind of measure.
4 *Improvements which add flexibility to the farm operation and improve long-term productivity.* Wood lots and agroforestry blocks, improved water supply systems, new fences.
5 *Improvements which enhance the farm as an attractive working environment.*

Budget

On the demonstration farms, after ranking works in order of priority with the farmers, John Marriott compiled a detailed costing of all proposed works in November, December and January before the main season for works from May to October. These costings were discussed with each farmer, and the contribution which the project budget could make on each property was outlined, leaving the farmer to determine the extent of works requiring his own contribution. The fact that the farmers' contribution increased proportionately as the project developed reflected their increasing confidence in the plans and their growing ability to implement them, although good seasons and steadily increasing wool prices undoubtedly helped.

There should be provision in all farm budgets for farm improvement works, for maintaining the quality of the land, just as people provide for the maintenance and depreciation of a car or house. The allocation made to this item determines the time scale for implementing the plan.

When priorities and physical and financial capacities have been determined, it is possible to mark the year in which each operation will be carried out or to prepare a third overlay illustrating the sequence of operations. This enables the landholder to look ahead, ensuring that adequate preparation is carried out before proposed operations.

For the Potter Farmland Plan, requiring intensive plan implementation on fifteen farms at once, it was necessary to plan, document and cost each operation well in advance in order to prepare orders for materials and seek quotes well before operations began—usually at least six months before. To assist in this process, I prepared a network diagram each year to illustrate the interdependency of the various tasks involved in carrying out the works programme, and the relationship between the works programme and other aspects of the project such as administration and public relations.

REVIEW

No farm plan is ever final and it is unlikely that first ideas will always be the best. When a first draft of the plan is completed, including a fully costed schedule of works, the plan should be discussed with the family and others involved in running the property. The Potter Plan farmers said that the involvement of the family, particularly children, in the planning process and implementation of the plans was one of the major advantages of the process. The first drafts of the plans were reviewed during and after the first year of works, and we found that the farmers played a more active role in the review process as the project developed.

The farmers do not consider that the whole farm plans are completed— they are continuing to evolve and will undoubtedly change with the farmers' changing life goals, different markets and family situations. Nevertheless, as the farmers progressively implemented their plans, the benefits of the work became more obvious, and their confidence and skill increased. Plans were inevitably refined and modified along the way, but as the basic planning steps outlined above were followed, such changes complemented early farm improvements, rather than being frustrated by them.

The process described above was carried out on all of the fifteen Potter Farmland Plan farms, but the extent to which the participating farmers were involved in the planning process varied. How much of a plan was implemented from 1985 to 1987 varied according to the demonstration value of the various farms, and the involvement of the farmers varied accordingly.

We found that an accelerated programme of works also accelerates the farmer's commitment to the planning process.

The basis of the project's approach is that the most appropriate person to prepare farm plans is the farmer, not government extension advisers. The farmer must 'own' the plan, rather than having it imposed from outside. As mentioned earlier, there are two key determinants of the worth of the plan—the impact it has on the farmer's thinking and knowledge, and the consequent impact its implementation has on the landscape.

(4)

Plans
into Action

● IMPLEMENTING THE PLANS—WHAT WE DID

I T IS VERY IMPORTANT to distinguish clearly between the whole farm planning process and the work on the demonstration farms. People often confuse the two, inferring that whole farm planning is about planting trees, shifting fences, laneways, planting out salt and fencing off dams. This is obviously not the case. If the project had been in a cropping region, for example, we would have been concerned with rotations, tillage practices, building soil fertility and organic matter and increasing biological and economic diversity. In rangelands, the outcome of whole farm or property planning would be changed to property layout, protection of water courses and strategic natural vegetation and careful consideration of watering points, and the plan would also concentrate on management strategies for normal variations in rainfall, stocking rates and supplementary feeding policies.

This chapter outlines the practical 'nuts and bolts' of implementing whole farm plans on grazing properties in a 600–700 millimetre winter rainfall area of south-western Victoria. The tables and graphs which follow summarise the work carried out on each demonstration farm in each year of the Potter Farmland Plan.

Summary of Works by Farm and Year, Glenthompson Area

Farm	Year	Fences km	Shelter belt km	Shelter belt trees	Clump trees	Single trees	Direct seeding (ha)	Wood lot ha	Pasture ha	Water
'Ballantrae'	1985	1.252	0.688	779	100	54	1.53			
	1986	2.744	0.824	905	1270		0.75			
	1987	1.430	1.430	1188	1000	54				
	Total	5.426	2.942	2872	2370	108	2.28			
'Cherrymount'	1985	1.289	1.021	1044	2000	54	0.08			
	1986	1.660	0.828	344	3700					
	1987	1.090	1.090	848						
	Total	4.039	2.939	2236	5700	54	0.08	0.00		
'Daryn Rise'	1985	3.962	1.587	1046	1280	152	0.06		11.25	
	1986	2.481	1.784	1785		72				
	1987	2.771	1.156	1260	536				7.70	2 dams
	Total	9.214	4.527	4091	1816	224	0.06	0.00	18.95	
'Fernleigh'	1985	2.448	0.423	423	944	100	1.00			
	1986	3.220	1.340	1685						
	1987	2.982	0.930	930	3370	100				
	Total	8.650	2.693	3038	4314	200	1.00	0.00		
'Nareeb Nareeb'	1985	2.522	1.462	1445	320	100	0.25			
	1986	1.680	0.800	800	260	108				
	1987	0.404	0.404	530						
	Total	4.606	2.666	2775	580	208	0.25			
'Reedy Creek'	1985	5.707	2.701	2592	2863	180	3.08	3.50		
	1986	3.163	1.757	2232	4010	72	0.20			
	1987	9.801	5.115	5115	1250				38.50	
	Total	18.671	9.573	9939	8123	252	3.28	3.50	38.50	
Total	1985	17.180	7.882	7329	5187	540	5.92	3.50	11.25	
	1986	14.948	7.333	7751	7600	244	1.03	0.00	0.00	
	1987	18.478	10.125	9871	10116	262	0.00	0.00	46.20	
	Total	50.606	25.340	24951	22903	1046	6.95	3.50	57.45	2 dams

Summary of Works by Farm and Year, Melville Forest Area

Farm	Year	Fences km	Shelter belt km	Shelter belt trees	Clump trees	Single trees	Direct seeding (ha)	Wood lot ha	Pasture ha	Water
'Helm View'	1985	8.571	2.135	1713	437	108	9.32		4.80	
	1986	7.320	3.057	2700	2750	90	1.25		36.00	5 troughs
	1987	9.868	3.194	3840	2580	198	1.80			
	Total	25.759	8.386	8253	5767		12.37	0.00	40.80	
'Pine Grove'	1985	4.496	0.640	270	500	30	1.65		4.30	
	1986	2.190	0.340	306			0.74			
	1987	1.292	1.457	1985		144				
	Total	7.978	2.437	2561	500	174	2.39	0.00	4.30	
'Warooka'	1985	2.828	1.216	1124			4.27			
	1986	1.721	0.890	880		36				
	1987	2.212	0.550	501	1290	36				
	Total	6.761	2.656	2505	1290	72	4.27	0.00	0.00	
'Willandra'	1985	5.719	1.934	1822	200	30	4.70		3.50	
	1986	3.460	1.342	1224	250	72	1.16			1 tank
	1987	15.047	1.594	2806		36	4.00		7.70	5 troughs
	Total	24.226	4.870	5852	450	138	9.86	0.00	11.20	
'Wyola'	1985	1.593	1.182	940		90	0.40		5.00	1 dam
	1986	1.360			340	36			7.50	5 troughs
	1987	1.919	0.775	944	264					
	Total	4.872	1.957	1884	604	126	0.40	0.00	12.50	
Total	1985	23.207	7.107	5869	1137	258	20.34	0.00	12.60	
	1986	16.051	5.629	5110	3340	144	3.15	0.00	5.00	
	1987	30.338	7.570	10076	4134	306	5.80	0.00	51.20	
	Total	69.596	20.306	21055	8611	708	29.29	0.00	68.80	

Summary of Works by Farm and Year, Wando Vale Area

Farm	Year	Fences km	Shelter belt km	Shelter belt trees	Clump trees	Single trees	Direct seeding (ha)	Wood lot ha	Water
'Barnaby'	1985	1.446	1.446	1370					
	1986	2.300			1085	72	0.20		
	1987	2.236	0.728	700	1100				
	Total	5.982	2.174	2070	2185	72	0.20	0.00	
'Benowrie'	1985	0.910				72	1.85		
	1986	1.000					0.40		
	1987	0.380	0.320	321		54			
	Total	2.290	0.320	321	0	126	2.25	0.00	
'Gheringap'	1985	3.526	1.627	1335		50			hydraulic ram
	1986	2.750	0.810	603			3.20		1 tank
	1987	7.297	2.764	2520	1220	36			4 troughs
	Total	13.573	5.201	4458	1220	86	3.20	0.00	
'Satimer'	1985	8.594	5.234	4397			3.20		
	1986	5.827	2.215	1715		144	2.02	5.50	3 dams
	1987	8.173	2.989	3613	2325				
	Total	22.594	10.438	9725	2325	144	5.22	5.50	
Total	1985	14.476	8.307	7102	0	122	5.05	0.00	
	1986	11.877	3.025	2318	1085	216	5.82	5.50	
	1987	18.086	6.801	7154	4645	90	0.00	0.00	
	Total	44.439	18.133	16574	5730	428	10.87	5.50	

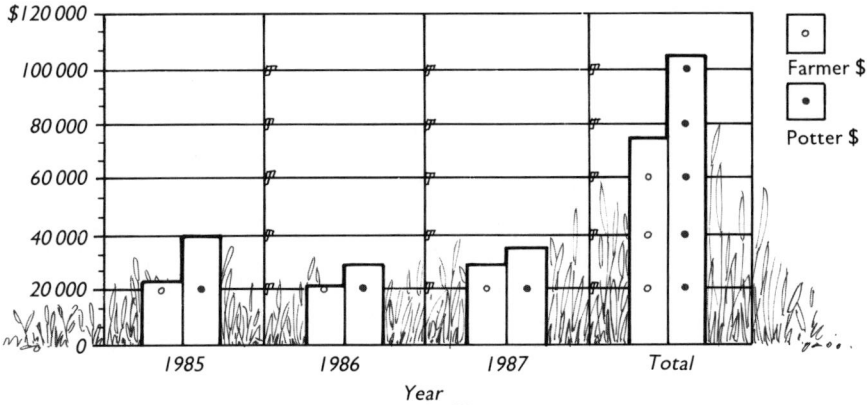

• Glenthompson works expenditure contributions, 1985-7

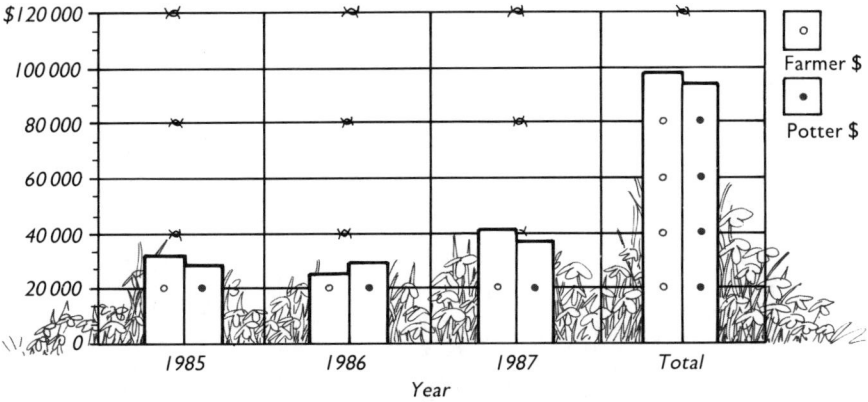

• Melville Forest works expenditure contributions, 1985-7

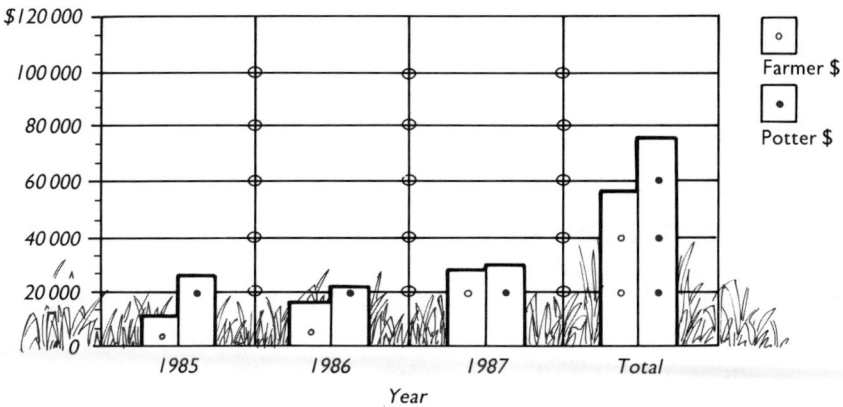

• Wando Vale works expenditure contributions, 1985-7

Summary of Works, Potter Farmland Plan				
	Year			
	1985	*1986*	*1987*	*Total*
Fences (km)	54.863	42.876	66.902	164.641
Shelter belts (km)	23.296	15.987	24.496	63.779
Shelter belts (trees)	20300	15179	27101	62580
Clump trees	6324	12025	18895	37244
Single trees	920	604	658	2182
Direct seeding (ha)	31.31	10.00	5.80	47.11
Wood lot (ha)	3.50	5.50	0.00	9.00
Pasture (ha)	23.85	5.00	97.40	126.25

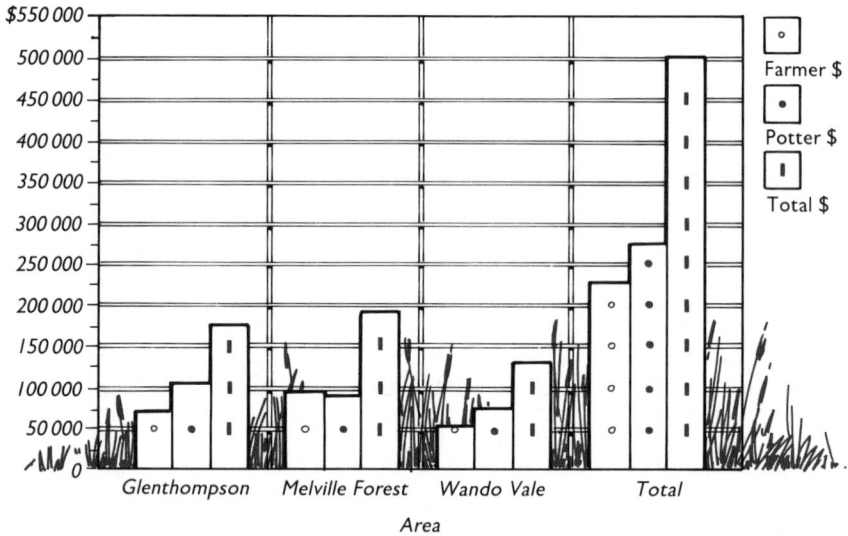

• Total works expenditure contributions by area, 1985–7

● HOW WE DID IT

The information in this section was distilled from a number of sources—
the practical experience of project staff, advice from specialists such as Bob
Piesse and Bill Middleton and, above all, the 'hands-on' innovation and
resourcefulness of the farmers themselves.

PASTURES

On 'Wyola', 'Helm View' and 'Willandra', low-lying areas were suffering from a complete lack of improved-pasture species and pasture composition had reverted to Sea Barley Grass, Buck's Horn Plantain and Yellow Buttons. In the drainage lines, areas of bare ground were increasing in size and being eroded by stock traffic, particularly during winter. The water trickling through these drainage lines varied in its salt content from 1800 to 2000 parts per million in spring to 5000 ppm at the end of summer. Bruce Milne and Peter Waldron estimated that these areas had a carrying capacity of one to three dry sheep equivalents (that is, they produced enough feed to sustain one to three Merino wethers) per hectare in 1985, although this is difficult to establish with accuracy as these areas were not separate grazing units.

The first impact of the whole farm planning approach was to fence off these low-lying, salt-affected pastures into separate paddocks. This enabled them to be treated as distinct land types, independently of the land types next to them, which were usually characterised by better drainage and more productive pastures in which clovers and perennial ryegrass persisted. After fencing out, these small paddocks were sown to salt-tolerant pasture species.

We asked the Hamilton offices of DARA and CFL which were the best species to sow and how best to establish them. Both departments hesitated to make recommendations, saying that they had little direct experience of establishing pastures on saline land in the region. However Peter Dixon recommended a pasture mix which had been used successfully at Neil Lawrance's property at Gatum: Tall Wheat Grass, Demeter Fescue, Palestine Strawberry Clover and Australian Phalaris.

The sowing rates and establishment techniques for this pasture are outlined in more detail in a paper presented by Bruce and Andrew Milne to the annual conference of the Grassland Society of Victoria in Melbourne on 18 June 1987, entitled 'A Farm Plan to Combat Dry-land Salinity with Pastures and Trees'. This presentation was an outstanding success which reflected the Milne brothers' excitement about the success of pasture establishment on 'Helm View' and production gains on formerly degrading land which had hitherto contributed only management problems of uneven grazing pressures, bogging and poor access.

Andrew Milne used the 'Helm View' Connor Shea coulter coil-tine seeder fitted with Baker points to direct-drill the same pasture species for Peter Waldron on 'Willandra' in September 1985. Within a year Peter was recording similar increases in carrying capacity to the Milnes. It was interesting that different establishment techniques were used at the two demonstration farms, as the Milnes used a conventional establishment technique of autumn cultivation, winter fallow and seeding into a fine-tilled seedbed in the spring. Both methods were very successful, but in 1985 we were lucky with the weather, which favoured spring sowing since we had dry conditions in September and October which allowed machinery access to work, followed by ideal conditions for germination in December and early January, with more than 125 mm of rain in December.

This early success provided insights into suitable species and establishment techniques to increase productivity in saline areas. By March 1986, the Milnes could graze weaners on their newly established pastures, despite the fact that the conventional wisdom suggested that Tall Wheat Grass should not be grazed for two years after sowing. The initial response on Peter Waldron's sites was not as spectacular, but Peter was grazing his new pasture about a year after sowing.

It is difficult to establish pastures in low-lying salt affected areas even in good seasons. The Potter Farmland Plan salt-tolerant pasture establishment programme was expanded significantly in autumn 1987, when over 100 hectares on 'Daryn Rise', 'Reedy Creek', 'Wyola', 'Pine Grove', 'Helm View' and 'Willandra' were seeded. Conditions in 1987 were less favourable for direct-drilling of pastures in autumn than they had been in 1985 and sites were unsuccessful, according to the number of seedlings per square metre preferred by pasture specialist Peter Shroder of DARA, Hamilton. Nevertheless, many of these areas have significantly increased carrying capacity compared with before sowing, and the fact that they are fenced into discrete management units enables them to be managed according to their needs, so that they still have the potential for optimum productivity.

FENCING

The fencing on the demonstration farms reflects the diversity of management styles of the farmers involved. More than 150 kilometres of fencing was constructed during the first three years of the project. In 1985, farmers were asked to specify their preferred fence designs and these various designs were implemented on the respective farms. However, because fencing represented a high proportion of the works budget, John Marriott and I looked carefully at designs and specifications and discussed fence design and labour costs with local fencing contractors and merchandisers. We adopted standard specifications for electric and non-electric (conventional) fencing, using both prefabricated and plain wire, and we organised three fencing field days between 1985 and 1987 at which speakers outlined the principles and practice of effective fencing and demonstrated modern techniques.

After the first field day, our experience with materials and labour costs for various fence designs and the chance to assess various fences in action, we felt confident enough to specify acceptable standards for both conventional and electric fences. If the Potter Farmland Plan farmers preferred a fence design which exceeded these specifications, it was agreed that the project would contribute up to two-thirds of the total cost of a 'standard' fence; in other words, landholders paid for any extra costs of building non-standard fences.

End Assemblies

End assemblies were critical to the effectiveness, reliability and longevity of fences. If they are well designed and built and if high-tensile wire, which can move freely against posts, is used, line posts then are only needed to keep wires off the ground, so they can be less substantial and further apart, thus reducing the materials and labour cost of the fence. A very effective end assembly for the most difficult conditions is shown in the diagram below:

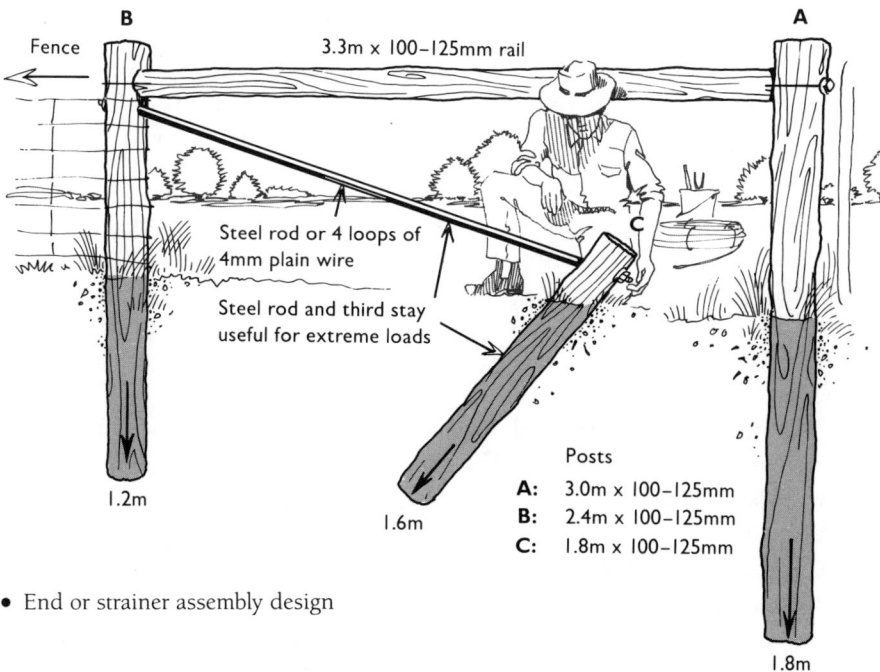

B

Fence

3.3m x 100–125mm rail

A

Steel rod or 4 loops of 4mm plain wire

Steel rod and third stay useful for extreme loads

C

1.2m

1.6m

Posts

A: 3.0m x 100–125mm
B: 2.4m x 100–125mm
C: 1.8m x 100–125mm

1.8m

• End or strainer assembly design

On the demonstration farms it was possible to construct adequate end assemblies without using the post marked 'C' in the diagram, and in most soil types posts were not driven as deep, but the design above will withstand very high overturning forces in highly plastic soils. Four engineering principles were followed for all the Potter fences and are worth outlining:

1   The depth of posts in the ground (depth of set) is more critical than post diameter in resisting overturning force, as long as posts are sufficiently strong not to break. It has been shown that increasing the depth of set from 0.75m to 0.9m (about 6 inches deeper) more than doubles the load which can be resisted. It can also reduce horizontal movement by 50 per cent and vertical movement by 33 per cent under the same conditions (load of 13 kN).

2   Driven posts are more effective than posts placed in a post-hole and back-filled with rammed earth. In one test, driven posts carried 50 per cent more load. Under the same load conditions (load of 13 kN), driven posts exhibited 60 per cent less horizontal movement and 80 per cent less vertical movement. Care should be taken to prevent water from moving down beside the post if cracks do emerge.

3   Horizontally stayed or box-strainer assemblies are about 25 per cent more efficient than diagonally stayed assemblies, provided rails are at least 3.3 metres long and the guy member maintains an adequate tension.

4   The length of horizontal stay (rail) is important. The advantage in added resistance to overturning forces between an assembly with a 2.4 metre rail and one with a 3.3 metre rail, as in the diagram above (each with unpointed posts driven 1 metre into the ground), is 64 per cent.

On all of the Potter farms where conventional fencing was used, strainer posts were at least 2.4 metres by 100–125 millimetres, rails were at least 3 metres by 100–125 millimetres and box-type end assemblies, rather than diagonal stays, were used.

Gates

As for fences, gate design followed the farmers' preferences, and farmers paid for the extra cost of any gates which exceeded the standard 3.6-metre, galvanised gate specifications. However, the Potter Farmland Plan created a particular need for gates into tree plantations and it soon became obvious to project staff that the farmers were not considering the need for access into trees carefully enough. Multiple-row shelter belts and wood lots needed easy entry for maintenance and management. Although access points did not need to be conventional galvanised steel swing gates, nor could they be 'cocky's' gates—since any fence is only as good as its weakest panel. It is a waste of time to put up an eight-wire fence with two barbs, three posts and nine droppers to the chain, if the gate in one corner is a length of slack 'Ringlock'. Gates into plantations are used rarely, so it is more important that they are effective as barriers than easy to open and shut.

A design for an opening into a block of trees is shown below.

An easy way to make an opening into wood lots or shelter belts is to attach the wires to droppers for two or three panels, rather than securing them directly to posts. The droppers are attached to the posts so that they can hinge at the bottom, using 4mm soft wire, and the section of the fence can be lowered and held down with pegs or weights, allowing vehicle or stock access.

This technique can be used with plain wire or fabricated fencing. It works better if the 'lay-down' panels are not at the strainer posts and if wires are able to move freely, rather than being secured tightly to posts.

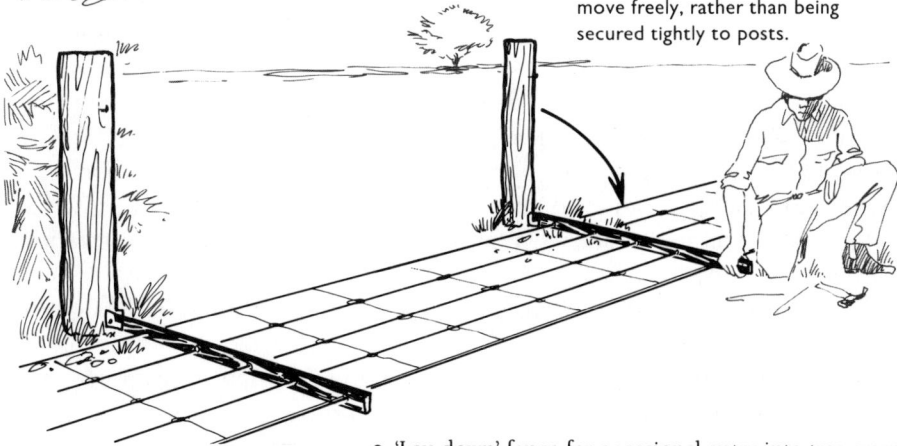

• 'Lay-down' fence for occasional entry into tree areas

This system worked well, even when wires were very tight, and it was very simple to install and use. It was used very effectively on 'Fernleigh',

'Ballantrae', 'Helm View', 'Willandra', 'Gheringap' and 'Satimer', in both electric and conventional fences. The only expense was the addition of two droppers and some tie-down pegs, and it avoided untying wires (and subsequent re-straining) or installing extra gateposts. It was easily adapted for electric fencing by using cut-out switches at either end.

Several of the farmers had been using 'lift-up' gates for many years before their involvement with the Potter Farmland Plan. We were keen to promote these gates, which are inexpensive, fast to erect, provide a very large opening and can be moved quickly if, for example, gateways become boggy. Offset lift or aerial gates such as the one below (but without the hinged top section, which is a recent innovation of farmers north of Cavendish), have been in use on 'Helm View' and 'Warooka' for many years, and have been used very effectively by the demonstration farmers.

Electric Options

Modern electric fencing has improved dramatically in design, equipment and construction techniques over the last twenty years. For less than half the cost of a conventional fence, well-designed and constructed electric fences offer better protection and increased flexibility of design. The technology, equipment, support services and expertise of contractors and merchandisers supplying electric fencing has improved tremendously over the last fifteen years. The Potter Farmland Plan took advantage of these developments and some of the powered fencing on the demonstration farms, notably 'Helm View', but also 'Satimer', 'Ballantrae' and 'Willandra', has been described as being amongst the best whole farm examples in Australia in terms of layout, specification and construction.

Electric fences are not only cheaper and faster to erect than conventional fences, but if they are well designed, constructed and managed they are also more effective. They are particularly useful in difficult situations such as for controlling vermin, fencing in very steep or boggy terrain, intensive grazing management (such as rotational grazing or agroforestry) or fencing in unusual shapes.

This section outlines the basic principles, and some useful design and construction hints, of tree protection as it was developed on the demonstration farms, but it is not intended to be a 'how to' manual of electric fencing. Much of the credit for the designs in this section is due to Bob Piesse, who has been a constant source of inspiration and advice to the staff and participants in the project since 1985, and to Bruce, Andrew and John Milne, who have demonstrated the potential of electric systems beyond doubt.

The most important stage in installing a power fencing system is planning. One of the most common complaints against power fencing is, 'I tried it

• To avoid stock damage to the creek, a 4500-cubic-metre dam was built on 'Wyola', with a pipe through the bank to reticulate by gravity to four nearby paddocks. The photograph at left shows the newly constructed dam and the aerial shot on the right shows the dam in March 1990, fenced out and revegetated to avoid sedimentation of the dam and thus improve the quality and security of the water supply. The native trees and shrub species seeded around the dam provide shelter and wildlife habitat, as well as protecting the dam overflow from erosion.

• It does not take long for wildlife to colonise fenced-off revegetation areas on farms. In 1986 our revegetation assessment crew placed small mammal traps in eighteen-month-old shelter belts on 'Satimer', which were quickly occupied by native rats and lizards. We employed students like Ian Bail and Caroline Hodges, pictured here, to evaluate and record the success of revegetation works from 1985 to 1987.

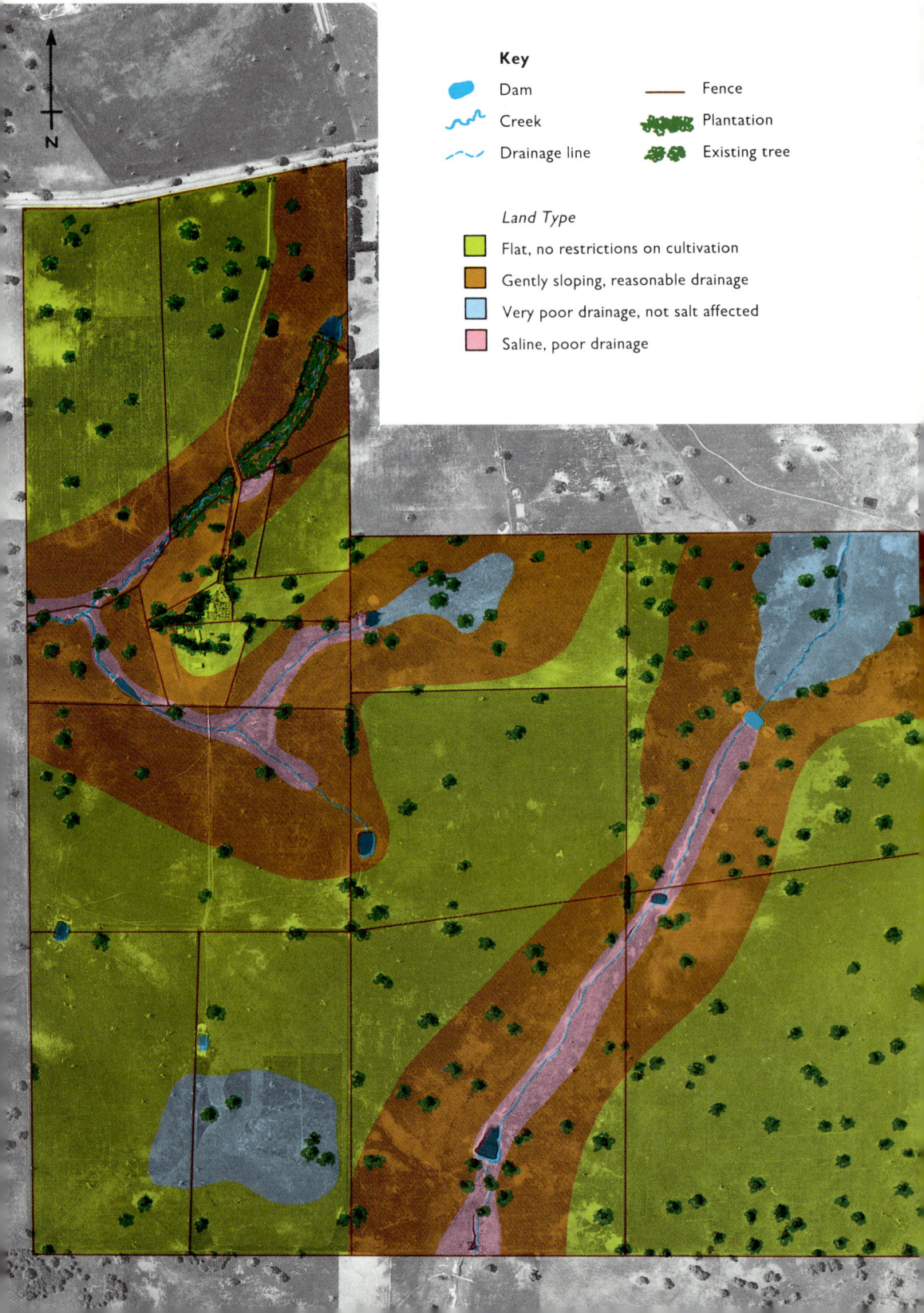

'Willandra'
Scale 1:13 000 (approximately)

• The base plan below shows land types and farm layout in 1984. Proposed changes and improvements are shown opposite. It may seem a lot of work, but Peter Waldron had put 90 per cent of this plan into operation within five years.

N

Key

Dam

Creek

Drainage line

Fence

Plantation

Existing tree

Land Type

Flat, no restrictions on cultivation

Gently sloping, reasonable drainage

Very poor drainage, not salt affected

Saline, poor drainage

'Willandra'
Scale 1:13 000 (approximately)

N

Direct seeding trial

Key

Fence
Laneway
Shelter belt
Dam
Creek
Drainage line
Wood lot
Octagonal regeneration plot
Individual tree in stockproof guard
Salt-tolerant perennial pasture
Windmill, Tank
Water pipeline and trough

• Attention to detail is critical for the successful establishment of trees on farms, whether planting by hand or machine. Note the good weed control along the ripline, the protection from rabbits with ultraviolet-treated plastic sleeves and the very efficient electric fence, with no strainers or stays needed.

Tangential view

Closed

3m

Elevation

50m

Open

Stock and vehicles pass under the
gate, which is effectively more than
20 metres wide. The offset section
is made of galvanised piping, hinged
at the base of the post. The gate
can easily be opened by one person
lifting the offset.

3m

**or
Open**

Vehicles higher than 3 metres drive
over the gate. Weights such as old
posts or tie-down pegs can be used
to lay the wires close to the ground
if a wide opening is needed.

Hinge point in
top stay of offset

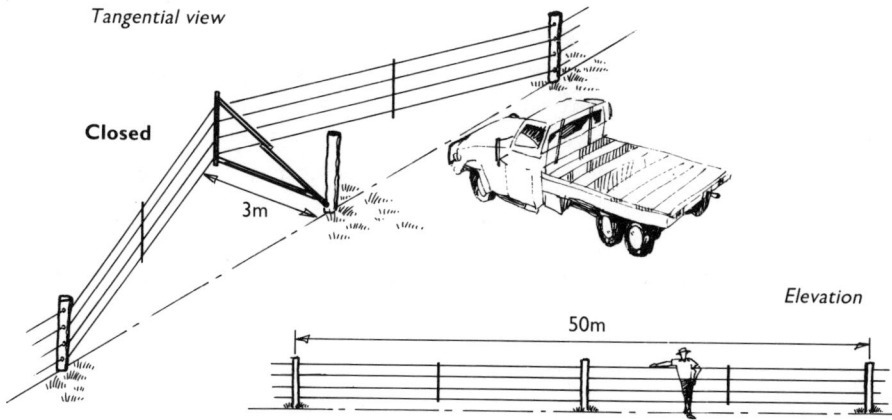

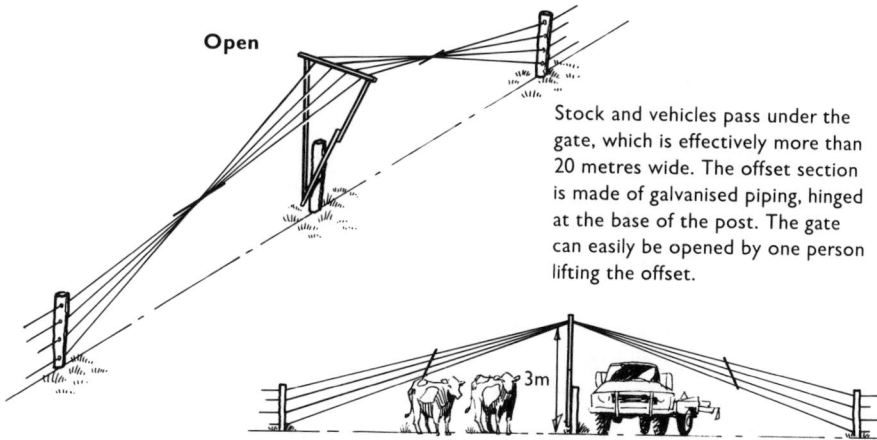

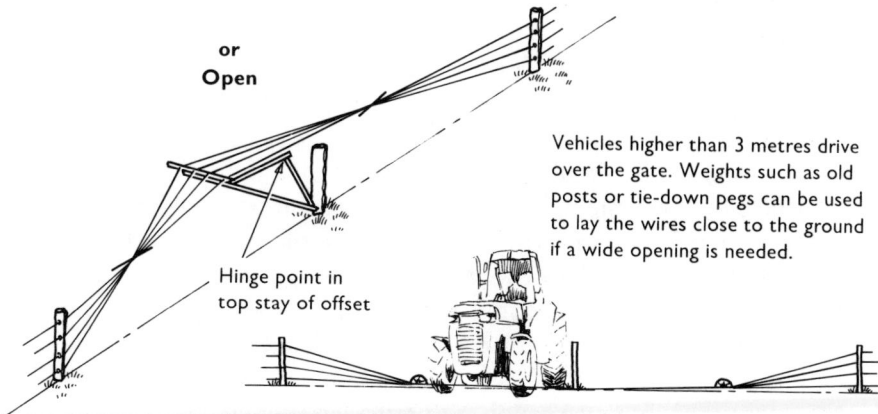

• 'Offset lift' or aerial gate, which can also lie down

years ago, but it didn't work so I've gone back to my old methods'. Usually the attempt involved trying to convert an existing old fence to an electric fence by adding an 'outrigger' wire, with poor energising, earthing, insulation and installation (particularly under gateways), no education of stock and little follow-up maintenance.

Ideally, power fencing systems should be designed from the ground up, rather than tacked on to existing systems, and the Potter Farmland Plan was an ideal opportunity for this to occur. We found that long-term maintenance was minimal if sufficient attention was paid to planning before installation.

Power Fencing Guidelines

Experience on the demonstration farms, in particular at 'Helm View', has generated the following general guidelines:

1   Energisers should be sufficiently powerful to allow for all present needs and any conceivable expansions, under the heaviest conductance challenge. However, if more power is required, it is preferable to use extra small energisers than larger central units, because of the danger of fires caused by the high energy generated by the very high-powered energisers which began to come onto the market in 1988.

2   Earthing should be very thorough, with earth points distributed regularly over the farm.

3   Fault finding should be an integral part of the system, with voltage alarms at each energiser, and a network of isolator switches, radiating from each energiser, with switches at each gateway, each tree belt and at other strategic locations to allow fast, accurate fault finding and easy maintenance. Routine use of the voltmeter or other means of testing in the course of normal farming operations takes very little extra time, saves on maintenance and greatly reduces the time spent finding faults.

4   It is essential to educate young stock to reduce stress on the system. Training yards or laneways (with fences with a greater number of wires and posts than those in the paddock—impossible to penetrate even if the power is off) should be used at weaning and off-shears to educate all stock to the system. Stock must be well accustomed to the fence before summer when power may have to be turned off on days of high fire danger. Before long every animal on the farm knows all about electricity, and fences become a psychological rather than a physical barrier. This means that they can be constructed using fewer wires and lighter posts further apart, with inexpensive gates and end assemblies.

5   Materials used in construction, in particular energisers, insulators, high tensile wire, insulated wire for under gateways, galvanised clamps,

isolator switches (cocos), spiral connectors, droppers and posts, should be the best available. Powered fences are much cheaper than conventional fences, but this is only an advantage if they are effective. The use of poor quality materials or rushed installation nearly always backfires and the natural advantage of the powered system is wasted.

6 Control of the distribution of power should be considered in the planning phase so that isolator switches are located not only to allow for quick fault finding but also for the isolation of particular sections without disconnecting the whole fence. For example, it is desirable to be able to turn off the powered wire closest to the ground if grass growth beneath the fence is high, either in fire danger periods or during winter when wet grass causes energy losses. Sensitive areas such as floodgates should also be capable of isolation without disrupting supply to the whole fence.

On the demonstration farms, we found that the requirements of powered fences designed to protect trees are the same as for conventional fences, but with careful attention to all the points mentioned above, powered fencing can meet these requirements more efficiently. It is even more important with powered fencing to leave sufficient space between outside tree rows and the fence for the tree to develop, as this lessens the likelihood of falling branches arcing between live and earth wires.

A Sample Fence

A fence which has been effective against sheep and cattle for several years on the demonstration farms is shown below:

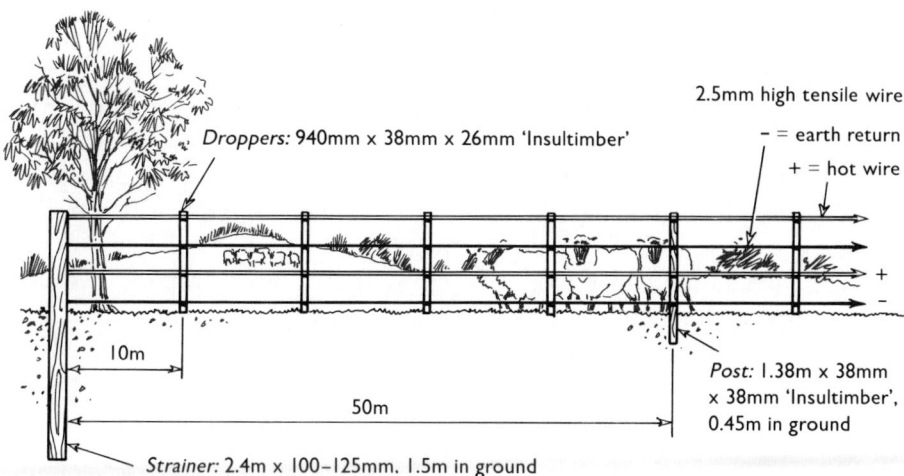

2.5mm high tensile wire

− = earth return

+ = hot wire

Droppers: 940mm x 38mm x 26mm 'Insultimber'

Post: 1.38m x 38mm x 38mm 'Insultimber', 0.45m in ground

10m

50m

Strainer: 2.4m x 100–125mm. 1.5m in ground

• Four-wire electric fence for sheep and cattle on flat ground

We constructed fences like this for $600–900 per kilometre, including labour costs. A conventional fence of equal effectiveness would have cost at least $2000 per kilometre, including labour. The Milne family, who have been using powered fencing for years and whose stock are educated to the system, have been able to use one- and two-wire fences to protect trees against sheep and cattle with complete confidence. One- and two-wire fences are effective on farms where all the stock are educated to the fence and the fence is rarely turned off, but they do not offer a visual barrier to stock, and they are less useful in areas where there are often days of extreme fire danger, although on such days stock are more likely to lie in the shade than to place pressure on the fence, especially if they are trained to electric fencing. The four-wire fence illustrated on page 91 is substantial enough to be an effective barrier to stock even when the electricity is switched off, such as on days of extreme fire danger.

The electric fencing on 'Helm View' is an outstanding example of a well-planned and carefully constructed grazing farm layout. John, Bruce and Andrew Milne were experts in electric fencing before their involvement in the Potter Farmland Plan and the electric fencing on 'Helm View' was of a very high standard when the project started. However the 25 kilometres of new fencing erected there between 1985 and 1987 increased the Milnes' already considerable experience.

Before the project their basic design was a six-wire fence with 75–100-millimetre by 1.8-metre pine posts every 10 metres, several droppers per panel, and box-section end assemblies. By 1987, the basic design had become a four-wire fence with one 100–125-millimetre by 2.4-metre pine post only at changes of direction, a 50-millimetre 'insultimber' post every 50 metres, and a dropper every 10 metres, with no end assemblies. This refinement of design and construction was perhaps accelerated by the Potter Farmland Plan, but it was led by the Milnes, with regular inspiration from Bob Piesse. John Marriott and I learnt about electric fence design and construction from the Milnes, rather than the other way around.

Gate construction is also more efficient with power fencing because gates are subject to less pressure from stock and, since less tension is required on wires, strainer assemblies do not need to be as substantial and hence as expensive as for conventional fencing. In fact, with properly planned farm layout and a network of laneways as on the demonstration farms, conventional gateways as we have known them for generations need no longer exist.

REVEGETATION

The project team planted approximately 110 000 trees and direct-seeded (applied tree seed directly to the site, rather than first growing seedlings which are then transplanted, as is the more usual method) 47 hectares of trees between 1985 and 1987. Almost all planting and seeding was done in spring, from late August through to mid-November. In 1985, approximately 31 000 trees were planted and 31 hectares seeded, 33 000 trees were planted and 10 hectares seeded in 1986 and in 1987 the planting programme expanded again to 47 000 trees of more than forty species, and seeding of 5.8 hectares. The organisation and implementation of a planting pro-gramme on this scale was complicated by the range of planting sites on the fifteen properties, the number of trees of each species for each row or portion of each site, the variation in seasonal conditions and site pre-paration, and the diversity of people doing the work. For John Marriott, the farmers and me, the two to three months of the planting season were an extremely busy time.

We had to work very long hours to carry out the works programme to budget and to schedule in 1985 and 1986 but, in retrospect, those years seem easy as by 1987 the fifteen participants in the project all wanted to accelerate their programmes and were willing to pay for the increase. The public relations and extension demands of the project were occupying most of my time, so John Marriott supervised the planting programme almost singlehandedly in 1987. This was a remarkable effort; 328 people from twenty-two different community groups were involved. These people were aged from five to seventy, many of them had never planted large numbers of trees before and most of the groups only wished to plant on weekends.

The graph overleaf shows that the overall survival rate of plantings was usually well over 90 per cent, with no follow-up watering or maintenance.

So how were these results achieved? They were achieved by following these planning principles (see Chapter 3):

1 Decide what role the trees are expected to perform.
2 Select the sites where the role(s) can be most appropriately achieved.
3 Select the species most appropriate to each site and the job in hand.
4 Prepare the site very thoroughly.
5 Plant or seed the trees with care.
6 Construct a protection system designed to be secure against the full range of possible hazards for the life of the tree community.

The sectors in the pie chart refer to
the proportional survival rates, after
1–2 years, of 236 separate planting
operations. The chart shows that
the vast majority of operations
achieved a survival rate of more
than 85%.

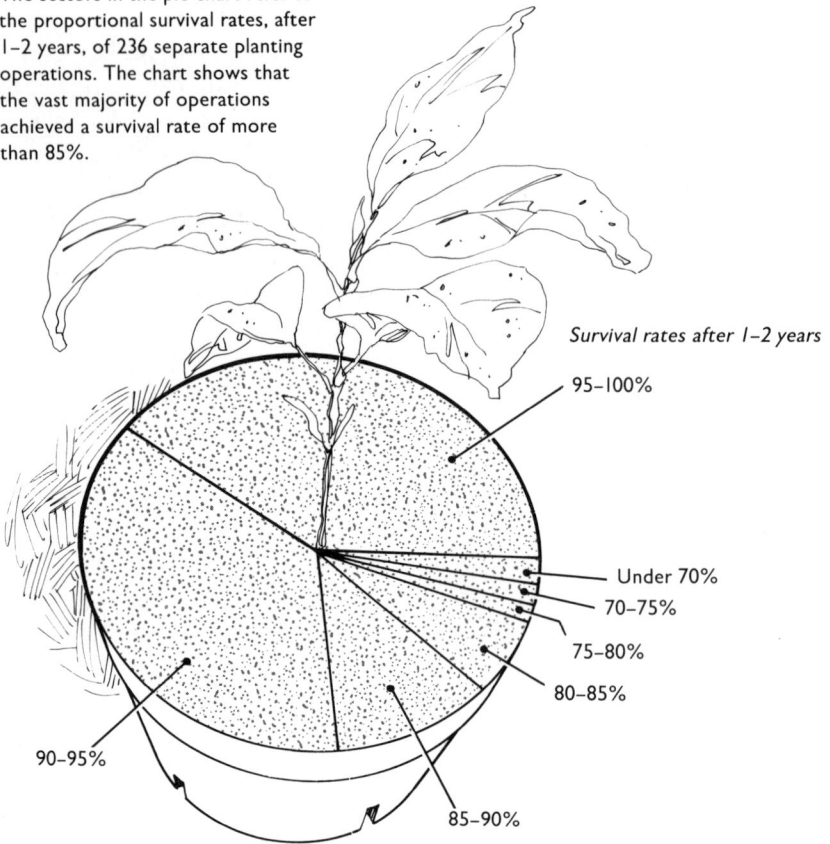

Survival rates after 1–2 years

95–100%

Under 70%

70–75%

75–80%

80–85%

90–95%

85–90%

• Summary of survival rates of Potter Farmland Plan plantings, 1985–6

Operations Planning

Before dealing with the physical task of planting trees, the exercise of allo-
cating, ordering and distributing seedlings is worth outlining. After 1985,
the demonstration farms were mostly planted with seedlings grown from
locally selected seed. Most of this seed was collected by the Cavendish crew
of CFL, although in 1985 crews from the Australian Trust for Conservation
Volunteers were also involved in seed collection. The main species collected
locally were Eucalyptus camaldulensis, viminalis, ovata and pauciflora; Allocasuarina verticillata; Casuarina muellereana; Bursaria spinosa; Banksia marginata;
Acacia melanoxylon, pycnantha, mearnsii and verniciflua; Melaleuca halmaturorum; and Leptospermum lanigerum, juniperinum and laevigatum. A portion
of this seed was kept for use in direct seeding, and the rest was sent to
the Vicflora nursery at Wail, near Dimboola, where it was grown into

seedlings. Although the Potter Farmland Plan had first priority on these seedlings, many of the main species and provenances were left over for others to buy, enabling them to purchase provenances adapted to the demonstration farm area.

The seedling requirement for each farm was finalised after operations plans were complete. However, for large orders such as the Wail order, nurseries need about one year's notice before delivery. For the Wail order a preliminary list of seedlings was sent in November of the year before planting, and the order was refined after operations plans were complete in mid-summer. The table overleaf shows the preliminary seedling requirement for the Melville Forest area in 1987. Similar tables were prepared for each area, using Microsoft Excel on the project's Apple Macintosh. This enabled easy alterations, simplified calculations of totals and facilitated print-outs of seedling requirements, by species, farm, area or nursery.

When the total seedling order and the requirements from each nursery were known, seedlings were allocated to each individual shelter belt, wood lot and gully, by species, provenance, numbers and nursery. These figures were entered on the paddock plans prepared by John Marriott. A seedling order for each farm was prepared like the one for 'Satimer' which appears on page 97.

After seedling orders were prepared for each farm, an order for each nursery (similar to the table overleaf, but with farms in columns across the top) was compiled and sent to the nurseries. The nurseries were asked to pack the seedlings in lots, farm by farm, with the name of the farm written on each box. Apart from Wail, the nurseries railed seedlings to Glen-thompson, Hamilton or Casterton, where they were picked up either by the landholders or by project staff. This system of seedling allocation, ordering and distribution evolved over three years, and worked to its potential in 1987. The use of a spreadsheet facilitated fine-tuning of the order and enabled more specific demands about packaging to be made of the nurseries, which made distribution a more foolproof operation.

When seedlings were delivered to the farm, the landholders were asked to keep them sheltered and watered until planting time. It soon became obvious that a short interval between delivery and planting was preferable to a long wait, as it is easy for farmers to forget to water seedlings. We suffered some losses through inattention at this stage. The paddock plans detailing the number of trees of each species for each ripline (the 400-millimetre deep furrows made with a tractor-drawn rabbit ripper to prepare soil for planting) were compiled by project staff. These were used in the field as working drawings. A sample paddock plan with revegetation details included appears on page 98.

Melville Forest Seedlings by Farm, 1987					
Species	'Pine Grove'	'Wyola'	'Warooka'	'Helm View'	'Willandra'	Total
Acacia howittii	0			19		19
A. iteaphylla	40		80	80	140	340
A. mearnsii	147	75	22	274	51	569
A. melanoxylon	207	140	126	372	276	1121
A. pycnantha	112	74	22	75		283
A. saligna				22		22
Banksia marginata	72	80		79		231
Bursaria spinosa	33	26	26	32	26	143
Casuarina stricta	154	125	46	98	277	700
C. cunninghamiana	123	30	23	400	70	646
C. glauca	69	50	265	685	151	1220
Eucalyptus bosistoana				35		35
E. camaldulensis						0
Albacutya provenance		40	260	376	80	756
Cavendish provenance	436	267	334	941	965	2943
E. cosmophylla		45	62	257	30	394
E. kitsoniana				148		148
E. leucoxylon rosea	137		62	98	177	474
E. maculata	75	134	83	25		317
E. occidentalis			80	418		498
E. ovata		45	333	466	186	1030
E. sideroxylon	523	167	83	79	153	1005
E. viminalis			40	12	50	102
Grevilleas					16	16
G. barkleyana				7		7
Hakea salicifolia				7		7
Leptospermum lanigerum					50	50
Melaleuca armillaris	50	50	144	287	36	567
M. decussata				215	36	251
M. ericifolia		40		166	36	242
M. halmaturorum			280	120		400
M. lanceolata			336	147		483
M. nesophilla				163		163
M. styphelioides				216		216
M. squarrosa				81		81
Myoporum insulare				48		48
Total	1921	1388	2707	6448	2806	14523

'Satimer' Seedlings by Nursery, 1987

Species	CFL Wail	NRCL	Just Trees	Meredith	CFL Creswick	Pete Smith	'Satimer' Total
Acacia decurrens				234		80	314
A. implexa			200				200
A. iteaphylla	40						40
A. mearnsii			90				90
A. melanoxylon	239			125			364
A. pycnantha				223	30		253
A. verniciflua	513						513
Banksia marginata	626						626
Bursaria spinosa	130						130
Casuarina stricta	504						504
C. cunninghamiana	26						26
C. glauca	150						150
Eucalyptus camaldulensis							0
Albacutya provenance	210						210
Pete Smith provenance						62	62
Edenhope provenance	264						264
E. cosmophylla	138						138
E. leucoxylon rosea	20						20
E. maculata					241		241
E. occidentalis			90				90
E. ovata		88			250		338
E. pauciflora	191						191
E. saligna	30				130		160
E. sideroxylon	17						17
E. viminalis	273						273
Grevilleas	224						224
Hakea suaveolens	100						100
Leptospermum lanigerum	100						100
Melaleuca armillaris	102						102
M. ericifolia	100						100
M. halmaturorum	98						98
Total	4095	88	380	582	651	142	5938
Price/seedling ($)	0.50	0.51	0.51	0.52	0.65	0.39	
Total cost ($)	2047.50	44.88	193.80	302.64	423.15	55.38	3067.35

Seed broadcast on 26 October 1987
Weather: warm with a light northerly wind
Seed:

Planting

170 g A. melanoxylon
165 g A. mearnsii   } boiled and soaked
271 g A. pycnantha
39 g E. camaldulensis
32.5 g C. stricta
13 g C. glauca
21 g M. halmaturorum
37 g E. viminalis
23 g E. ovata

TOTAL: 771.5 grams, mixed with chick crumbles

Because some of the acacia seeds had been boiled for a couple of days and left in a plastic bag, Peter also added:

150 g A. melanoxylon
150 g A. mearnsii
250 g A. pycnantha
30 g E. ovata
50 g E. camaldulensis
30 g C. glauca
15 g C. stricta

Fence →

Area planted

Lane

TOTAL: 675 grams

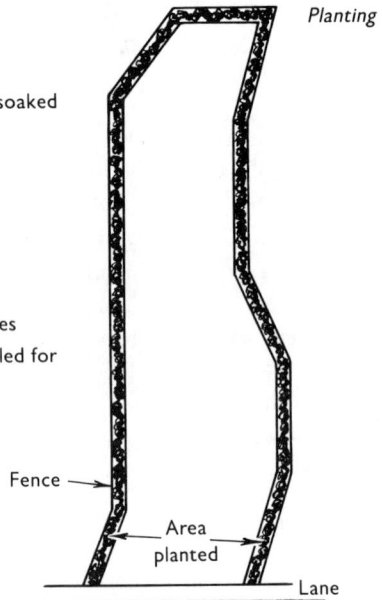

Pasture Establishment

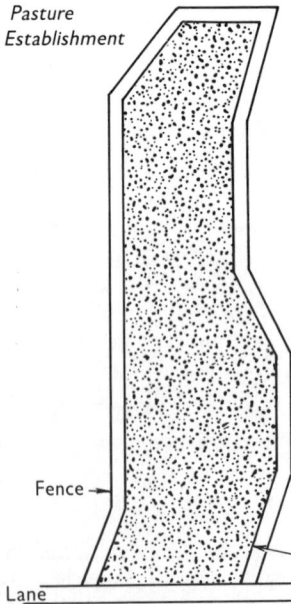

Fence →

Lane

Area sown to pasture

All this area received a split application of 1 litre sprayseed (per spray) with Agral and Lemat at 100 mL/h in second spray on 14 and 22 April 1987

Direction drilled by Connor Shea drill and Baker Boots on 24 and 25 April 1987

7 kg Tall Wheat Grass per ha at	$2.68
2.5 kg Sirosa Phalaris per ha at	$5.10
2.5 kg Seedmaster Phalaris per ha at	$3.54
4 kg Demeter Fescue per ha at	$3.33
10 kg Trikkala subclover (inoculated and lime coated) per ha at	$1.48
1.5 kg Strawberry Clover (inoculated and lime coated)	$5.64
120 kg Superphosphate per ha at	$150.00 per tonne

Total cost:	Seed	$71.58 per ha
	Super	$18.00 per ha
	TOTAL	$89.00 per ha

Pasture area was subsequently sprayed twice again with Lemat for red-legged earthmite control on 7 May 1988 and 12 June 1988 at 200 mL/ha each time.

• The paddock plan of 'Helm View', Melville Forest

Similar paddock plans were also prepared for fencing, site preparation and maintenance.

Shelter Design

A reappraisal of shelter planning is required in Australia. It is important to think carefully about the precise requirements for shelter, rather than rushing out and planting trees along fence lines. The obvious prerequisite for an effective shelter network is a whole farm plan, with management units sensibly defined according to land type and water supplies, access, fire protection and measures for land-degradation control considered and mapped out, if not implemented. In preparing whole farm plans, we had identified many shelter belt locations. Other areas with the potential to provide emergency shelter, such as gully plantings, wood lots and revegetated rocky areas were self-evident, before shelter *per se* was considered.

In preparing the specifications of the Potter Farmland Plan shelter belts, we were conscious of the need to improve the design of shelter belts and to demonstrate good design in the field. Past experience in designing shelter belts for farmers in central Victoria and in the Goulburn Valley, backed up by a study tour of New Zealand in June 1985 and discussions with Bill Middleton, Rob Youl, Steve Burke, Kevin Ritchie and Greg Wallace of CFL and Rod Bird and Keith Cumming of DARA, led me to re-appraise the recommended specifications for farm shelter belts.

One of the characteristic features of the Western District of Victoria, particularly on the basalt plain extending south-east of Hamilton to Geelong, are the hundreds of kilometres of mature shelter belts of cypress (*Cupressus macrocarpa* and *C. lambertiana*), pines (*Pinus radiata*) and Sugar Gum (*Eucalyptus cladocalyx*), established from the turn of the century through to the late 1930s. On many of the older stations these plantations are extensive, often two chains (or 40 metres) wide, with a laneway up the middle through which horses and drays used to travel.

However, very few of these plantations have been properly managed to ensure that they effectively fulfil their original purpose, which was to provide shelter for stock from cold, driving winds. In many cases the inside fence of the shelter belt has been allowed to deteriorate and has not been replaced, which has allowed stock to browse on the lower branches. When lower branches are browsed, the resultant gap between the foliage and the ground creates a venturi for the wind, which can be up to 20 per cent faster through the gap than out in the open paddock. Not only are windspeeds higher through the gaps, but stock tend to camp under the trees, which increases their risk of exposure to cold weather, although the trees do provide shade in hot weather. When stock camp under trees, patches of bare ground develop which tend to become muddy in winter and dusty in summer, which is a nuisance for wool growers, particularly fine-wool producers.

Maintenance of fencing is not the only problem of many existing shelter belts in Australia. More thought should be given to the design of shelter, not just by landholders, but by the staff of government extension services and nurseries and private consultants who advise them.

Trees provide shelter by absorbing energy from the wind as it rustles leaves on its way through, but many existing shelter belts, particularly of pines and cypress, have become so dense that the wind does not pass through them at all. Instead, it bounces off and passes in a turbulent stream over the top of the trees. These dense shelter belts provide excellent shelter for a distance of one to three tree heights on their lee side, which makes them useful around buildings, yards and very small holding paddocks, but of very little use in larger paddocks.

In open paddocks, where shelter for ten to twenty tree heights might be required, shelter belts should be permeable, so that daylight is visible through the trees when they are viewed from side on. Some of the wind must be able to penetrate through the trees, rather than deflecting over or around the shelter belt. Designing a permeable shelter belt requires forethought and a commitment to manage the trees to ensure that they remain permeable as they grow. It is unrealistic to expect to 'plant, fence and forget' a shelter belt and provide effective shelter for the fifteen tree heights that a permeable shelter belt can provide. To do so would require the trees to be planted far apart and for all of them to survive. As well, much patience would be needed, because it would be many years before the shelter belt could provide effective shelter. Rather, it is preferable to plant trees at such a spacing that they will provide shelter in the first five to ten years. This also allows for some losses and provides a more favourable environment for trees to develop with mutual support and protection and with competition for light causing increased height growth.

Standard Potter Farmland Plan Shelter Belt

The design of most of the shelter belts used in the PFP between 1985 and 1988 is illustrated on page 101.

This shelter belt design ensures that reasonable shelter close to the trees is provided within five years, but it allows for trees to be thinned to maintain permeability, providing farm timbers and some cash flow, when the shelter belt becomes too dense and tree growth is inhibited by competition. It also incorporates two rows of bushier, understorey species which are predominantly indigenous shrubs such as *Acacia melanoxylon*, *mearnsii* and *pycnantha*, *Allocasuarina verticillata*, *Casuarina muellereana*, *Banksia marginata*, *Bursaria*

Tall eucalypts to be thinned for
posts, poles and firewood at age
15–20 years. On Potter Farmland
Plan, usually *Eucalyptus
camaldulensis* and *E. maculata.*

Indigenous eucalypts such as
*Eucalyptus camaldulensis,
E. maculata, E. viminalis, E. ovata,
E. leucoxylon rosea.*

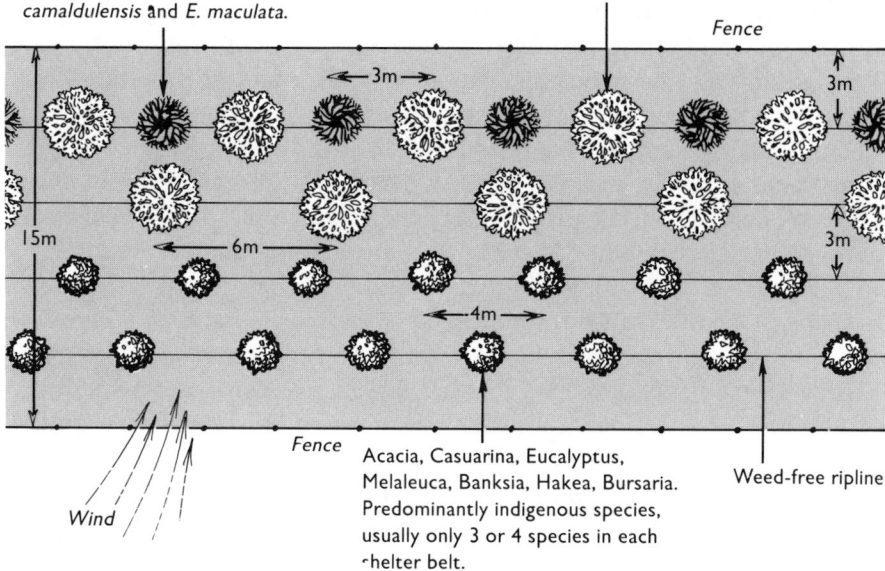

Fence

←3m→   3m

15m   ←6m→   3m

←4m→

Fence   Acacia, Casuarina, Eucalyptus,
Melaleuca, Banksia, Hakea, Bursaria.
Predominantly indigenous species,
usually only 3 or 4 species in each
shelter belt.

Wind

Weed-free ripline

• Design of standard Potter Farmland Plan four-row shelter belt

spinosa and *Hakea sericea*. These shrubs provide more than low shelter. They turn the shelter belt into a wildlife corridor, with the advantages of attracting local species, discussed in Chapter 3, with appropriate flowering times, nesting sites and nectar sources for migrating birds and insects, and the ability to persist in local conditions conferred by evolution over thousands of years.

The design above conveniently used 1000 trees per kilometre, which made calculation of seedling requirements straightforward, and it also allowed for a few losses, as most of the species planted have the ability to fill gaps through natural regeneration or through 'suckers'. They can also recover from fires, which was clearly demonstrated on many Western District properties after the 1977 fires. Shelter belts of pines and cypress were killed outright, whereas native species, although completely burnt, either recovered with vigorous new growth or regenerated prolifically from seed in the ashbed created by the fire. In fact many of the older native shelter belts were more effective several years after the fire than they had been before.

Some of the stock pressure on fences around wood lots and shelter belts was reduced on the demonstration farms by thoughtful planting. Preferably, tree rows were not planted within 3 metres of fences. This was important to allow trees room to grow and to keep their low branches unbrowsed, avoiding the wind tunnel problem which develops under many shelter belts where landholders have been reluctant to 'give up' too much ground. Having a 3-metre space between fences and trees also discourages stock from reaching over the fence, thus easing pressure on the top wires. An extra metre on either side of a shelter belt only takes up 0.2 hectares per kilometre of shelter belt, which is insignificant in comparison to the benefits of improved tree growth, wind management and tree protection. The 3-metre gap was used on both sides of almost all the demonstration farm shelter belts, and it allowed tractors and slashers to work between tree rows in the early years, to reduce the fire hazard of long grass between the trees.

These 15-metre wide shelter belts are big enough to serve as emergency shelter for sheep in extreme conditions. In such conditions, early recognition of the hazard and prompt action is critical, so the access points to the shelter belts are very important. Sections of fence which can be laid down, as explained in the diagram on page 87, are much more useful in such emergencies than the more expensive narrow gates.

Variations in Shelter Belt Design

We implemented several variations on the standard design. In some cases (where it was not intended that trees would be thinned), we planted the lower shrub species in the outside rows and tall species in the middle rows. These 15-metre-wide shelter belts are ideal for most situations, particularly on grazing farms where the use of land is not intensive. However for very small farms, or for small paddocks near sheds and yards, narrower shelter belts may be justified, although most of the Potter farmers now believe that wider shelter belts are the best in the long term, and their benefits more than outweigh the small amount of extra land they use.

David McDonald and Ross Kitchin expressed concern about the amount of land removed from grazing for shelter belts and, after consultation with project staff, narrower shelter belts were designed and demonstrated. In each case they were designed primarily to provide shelter (with some potential for fodder and timber production) using the minimum amount of land. They were not intended to be wildlife corridors or refuge areas for stock, nor were they designed to enhance the landscape (all though all of these were possible). It is worth examining each design in turn, against the principal objective of providing shelter.

Ross Kitchin needed shelter on the north and west sides of a small (about

2 hectares) holding paddock on a sandy, exposed tableland near his wool-shed and was reluctant to lose grazing land to trees in such a small area. The design of a shelter belt which was planted in spring 1985, and which is providing useful shelter and fodder already, is illustrated in the diagram below.

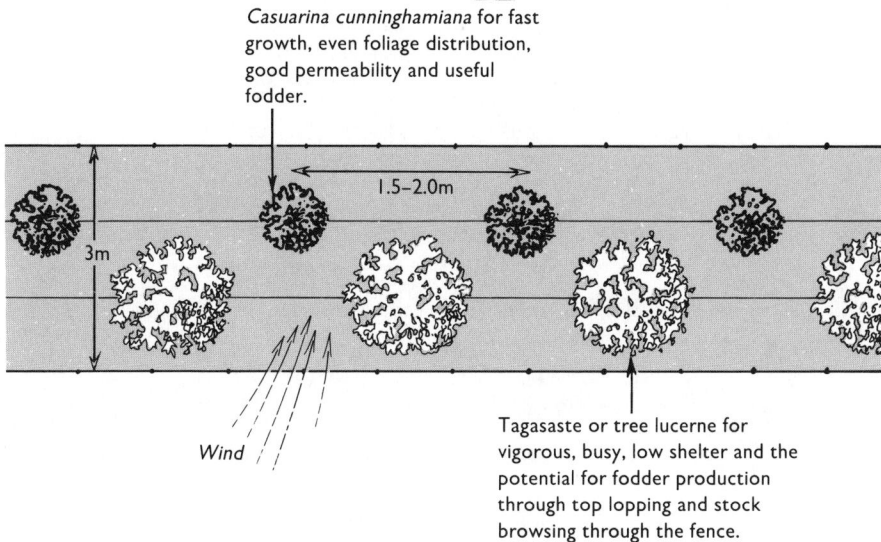

Casuarina cunninghamiana for fast growth, even foliage distribution, good permeability and useful fodder.

1.5–2.0m

3m

Wind

Tagasaste or tree lucerne for vigorous, busy, low shelter and the potential for fodder production through top lopping and stock browsing through the fence.

• Narrow, two-row shelter belt with fodder

Neither of the species planted in this shelter belt is indigenous, but they are fast-growing legumes which do grow very well in western Victoria, although Tagasaste requires very good drainage. River sheoak, *Casuarina cunninghamiana* is an excellent tree for shelter belts in the demonstration areas, as it grows uniformly well and tends to form thickets through suckering, which prevent gaps forming, a very useful characteristic in narrow one- or two-row shelter belts. It is important to avoid gaps, as wind speed increases through them and it is very difficult to re-establish a tree in a gap after the shelter belt reaches a height of more than 1 or 2 metres. In very sandy soils, such as near Esperance in Western Australia, wind erosion 'blow outs' can occur due to the turbulence and extra wind speed through gaps in shelter belts.

Russell and David McDonald had planted a two-row *Pinus radiata* shelter belt along the western side of a laneway on 'Daryn Rise' before their in-volvement in the Potter Farmland Plan. As a shelter belt it was not very successful, with a survival rate of less than 20 per cent. However it was in the right place for the right purpose, so we decided to improve it to provide an example of a carefully designed pine shelter belt, incorporating fodder and timber production. The design of this shelter belt is illustrated in the diagram below.

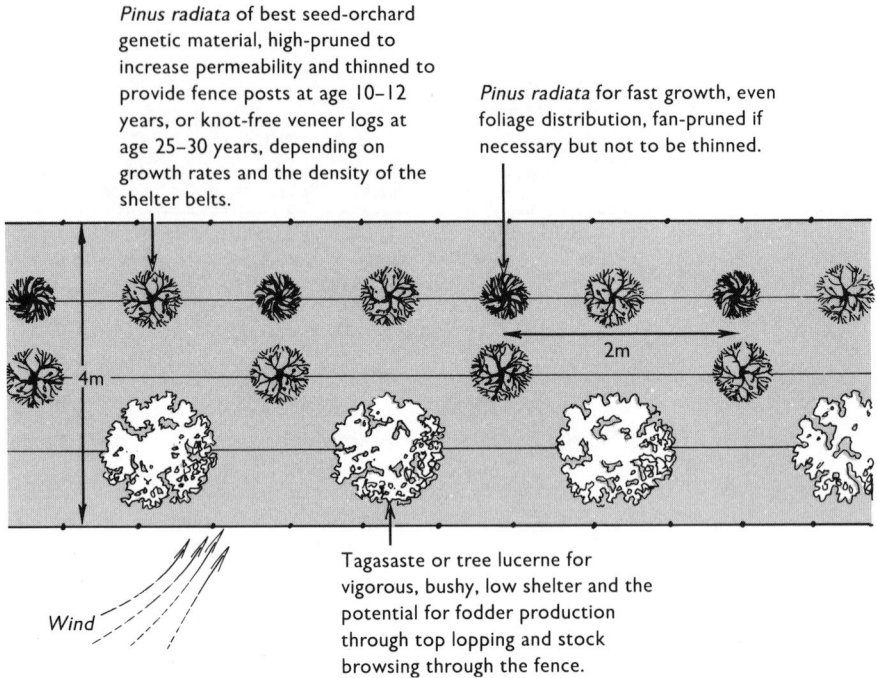

Pinus radiata of best seed-orchard genetic material, high-pruned to increase permeability and thinned to provide fence posts at age 10–12 years, or knot-free veneer logs at age 25–30 years, depending on growth rates and the density of the shelter belts.

Pinus radiata for fast growth, even foliage distribution, fan-pruned if necessary but not to be thinned.

2m

4m

Wind

Tagasaste or tree lucerne for vigorous, bushy, low shelter and the potential for fodder production through top lopping and stock browsing through the fence.

● Three-row shelter belt with fodder and timber

In New Zealand it is standard practice to prune the branches which stick out over the fence from cypress and pine shelter belts to allow more breeze through the tree, increasing the permeability and effectiveness of the shelter belt. Several innovative farmers have also high pruned every second tree, to produce high quality veneer logs. A notable example of such a shelter belt, 22 years old, was clearfelled in 1985 and yielded a return of more than NZ $30 000 per kilometre. Such growth rates are rarely achieved on farms in Australia, but the potential for post, rail and sawlog production from carefully managed shelter belts or 'timber belts' is great and the shelter they provide is usually superior to the 'brick-wall' effect generated by neglected pine and cypress shelter belts. The amount of work involved in pruning on the average farm would be several days per year at most, and CSIRO research in Western Australia has revealed that pine prunings are useful fodder for stock if they are eaten while still fresh.

• Before and after: it is difficult to overemphasise the improvement in stability and productivity which the whole-farm plan has conferred on this section of 'Willandra'. Production has quadrupled, the spread of salt has halted and shelter and wildlife habitat have been provided, all at minimal cost. Note the efficient electric fencing and the success of direct seeding for cheap tree establishment. The aerial shot, looking back towards where the ground photographs were taken, illustrates the land-type fencing and the value of these formerly saline areas for summer and autumn grazing after they have been rehabilitated as part of the whole-farm plan.

• Before and after: taken four years apart, these photographs show the development of a typical Potter Farmland Plan shelter belt. While growth rates are not spectacular, survival rates have been excellent and as this belt is on the boundary between two land types, it has helped to improve the efficiency of grazing and consequent productivity already. Shelter belts must be designed in the context of a whole-farm plan.

In the McDonald example, Tagasaste was used to provide bushy, low shelter and fodder. Tagasaste is one of the few shrubs which will persist in close association with pines and cypresses, which are very hungry surface feeders, so it had the potential to play a useful role. However, it has not been as successful as expected at 'Daryn Rise' because the site, although on a high slope, is not well drained and the Tagasaste seems to be suffering from wet feet. We obtained the best bred pines possible, purchasing the latest generation of New Zealand Forest Service seed orchard stock through Colac Pines, a private softwood plantation company. Genetic gains in *Pinus radiata* in trees with a good 'pedigree' are significant and there are many advantages in terms of form, growth rate and avoidance of defects such as bends, sweeps, multiple leaders and 'speed wobbles' (the growth defects which occasionally affect pines grown on former pasture sites with a history of high fertiliser use). This is critical for farm timber belts. With only a few rows in most farm shelter belts and often long distances from markets, it is important that every tree is a winner if commercial returns are to be obtainable.

Mid-Paddock Shelter

Perimeter shelter belts were only one component of the Potter Farmland Plan shelter strategy. When paddock sizes increase beyond about 10 hectares (depending on the shape of the paddock), the area of effective shelter provided by perimeter shelter belts is limited. Consider, for example, a square paddock of 400 metres by 400 metres (16 hectares), with shelter belts on the southern and western fences designed to protect the paddock from prevailing winds. Assuming that the shelter belts are permeable, with an even height of 10 metres and a zone of effective shelter extending 15 tree heights or 150 metres into the paddock, they shelter less than 40 per cent of the paddock. Under set stocking regimes, stock get to know where the shelter is and will use it, but in severe conditions they tend to be forced by the wind away from the sheltered area.

We considered this issue early in the Potter Farmland Plan, and realised that it is better to provide shelter in the areas to which stock are forced in extreme conditions than to attempt to shelter the entire paddock. I doodled with designs for mid-paddock shelter belts and came up with a boomerang-shaped design which could be located where shelter is needed most. The rationale for choosing a bent design was to deflect wind around the apex of the shelter belt and to provide protection against wind from a greater range of directions than is possible with a linear shelter belt.

The diagram above illustrates the first boomerang-shaped mid-paddock shelter belts implemented on the Potter farms.

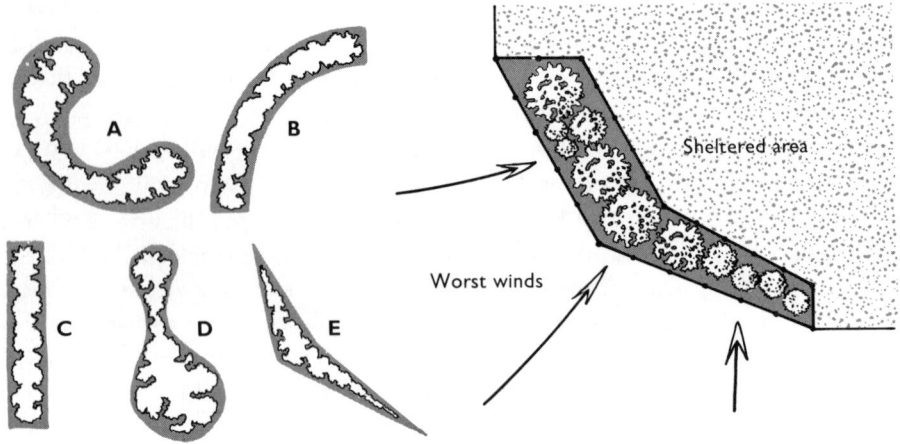

• Sample shapes of mid-paddock shelter belts

The top part of the diagram shows the location of a boomerang in the north-east corner of a paddock subject to cold winds in the sector from west to south. The dotted lines represent temporary fences which can be used to turn the sheltered area into a holding paddock, or to heavily graze an area in early summer for fire protection. The Potter farmers found it best to carefully measure such boomerangs when marking out the corner posts to ensure that the ends are parallel to the adjacent fencelines and that each end is the same distance from its corresponding fence, to avoid extra turning when the paddock is cultivated.

After successfully implementing several examples of the top design above on 'Reedy Creek', 'Fernleigh', 'Nareeb Nareeb', 'Helm View', 'Pine Grove' and 'Satimer', we also tried a range of other shapes, including designs A to E above. Designs A and D are in place on 'Cherrymount', 'Helm View' and 'Barnaby'. Their advantage was that they followed the contours of the land, their rounded outline minimised turbulence and there were no sharp corners for stock to track around. These shapes did not require strainers or stays, which kept materials costs down, but they were 'fiddlier' to fence and in some instances required longer posts.

Design B is in place on 'Ballantrae' and 'Satimer'. It is quite a simple design which works well provided attention is given to the ends. We made a mistake at 'Satimer', placing the curve so that the fence ran off the contour and slightly downhill. The sheep tracks leading down the slope around the end of the shelter belt are now a potential erosion hazard and may require surfacing with gravel at the corner to prevent washing.

Design C is on 'Daryn Rise', 'Fernleigh', 'Helm View', 'Willandra' and 'Pine

Grove'. It was very straightforward to construct, although expensive due to the number of strainers and stays, as was the first boomerang featured above, which was implemented on 'Nareeb Nareeb', 'Fernleigh', 'Daryn Rise', 'Cherrymount', 'Reedy Creek, 'Satimer' and 'Pine Grove'.

Design E is in place at 'Helm View'. This is a very simple way to create a boomerang as long as the ends are not on erodible soils. A big advantage of the electric fencing at 'Helm View' was that none of the designs above required strainers or stays, as there was no stock pressure on the 4 wire fences and corner posts were 2.4 metre posts driven in about 1.5 metres, which eliminated the need for stays.

In 1985 it appeared that 'boomerangs' were an innovation of the Potter Farmland Plan, but it is clear that farmers in the past have had the same idea. In driving around Victoria, I have since seen several thirty- or forty-year-old versions of the original bent design, one six- to eight-year-old version of design B and many examples up to fifty years old of design C. One farmer who has had a boomerang on his property for many years is convinced that it has saved shorn sheep on several occasions and Barry Levinson has noticed that ewes and lambs are using boomerangs before the trees are as high as the fences. However location is critical and careful observation of animal behaviour in severe conditions is required before locating mid-paddock shelter belts. They are more expensive to fence per tree than perimeter shelter belts and it is important to ensure that they achieve what they are designed to do.

Another innovation in planting design conceived by a farm-tree grower near Hamilton can also be used to provide emergency shelter. When cultivating or spraying a paddock, Jock Bromell makes a careful note of the shape of the 'awkward bit in the middle'. When the paddock is nearly finished and a lot of turning is involved to work out the centre, Jock puts in some marker pegs at the corners of the centre piece, regardless of its shape. He then fences out that (one half to two hectares) section and plants it to trees!

Another aspect of mid-paddock shelter which deserves more consideration when farm revegetation is planned is shade. Many people plant trees in individual guards in an attempt to imitate the park-like landscape which is achieved with majestic old trees dotted in paddocks of lush green pasture. The difficulty with single trees is that the sole tree established within the guard is exposed to all the stresses which are afflicting lone trees in many of our agricultural areas, in an environment totally different from that in which the much-admired older trees grew up.

There are two alternatives to the planting of individual shade trees which should both be investigated before making the investment in individual guards. The first option, which most people seem to have overlooked in many areas, is to make some effort to prolong the life of the paddock trees we have left and to enhance them where possible with natural regeneration and the establishment of understorey species.

We experimented with possibilities for 'renovating' deteriorating Red Gums at Wando Vale on 'Benowrie'. At the suggestion of Bill Middleton, six old Red Gums which were in various stages of crown dieback were pruned severely, using a hired 'cherrypicker' to remove dead or dying branches, thus concentrating the resources of the tree on producing fresh growth. Results to date are very encouraging, as the pruned trees responded well, and it seems their life has been prolonged. An interesting sidelight is that the cherrypicker cost $300 to hire for the day, but 6 kilograms of seed, which had a market value of $100 per kilogram, were collected from one tree! We also established clumps of regeneration enhanced by seeding of understorey acacias, melaleucas and leptospermums. The pruning technique is not regarded as a cure for tree decline, merely as a means of prolonging the life of trees suffering from dieback. The long-term measures, demonstrated at 'Benowrie', to encourage the development of diverse communities of indigenous trees and shrubs, offer the only hope of retaining the majestic character of the Red Gum woodlands across the Dundas Tablelands from Casterton to the Grampians.

The second alternative to individual shade trees, which usually makes more sense if there is sufficient space available, is to establish trees in small clumps, containing trees and understorey shrubs in a mutually supportive environment. Clumps are ecologically more stable: they are likely to remain healthy for much longer than single trees because they provide better habitat for natural predators of insect pests, they are better protected from wind and they allow for natural genetic variation. As well as their value in attracting birds and other wildlife, clumps tend to look more natural when dotted around the farm. This point was made very clearly by Ione Levinson, who rejected the notion of 200-litre drum, single-tree guards for aesthetic reasons, saying that she did not want 'Reedy Creek' to 'look like a Shell depot'.

In search of a more cost-effective and ecologically sound means of providing mid-paddock shade and shelter, and protection for patches of natural regeneration, we costed different shapes and specifications, and came up with an octagonal clump 50 metres in circumference, chosen simply to get four clumps per roll of 'Ringlock', which was 16 metres in diameter, with posts 6.25 metres apart. Theoretically such a clump can contain about thirty trees, about six of which are tall species which can be progressively thinned (often yielding wood products such as firewood or fence posts) eventually leaving one or two healthy big trees—and about twenty shrubs, for a protection cost of approximately $100 including labour. This compared very favourably indeed with $10 to $12 per sheep guard and $30 per cattle guard for individual trees. With single trees there are no second chances within each guard and even with a sound guard the result may not be satisfactory due to a poor seedling or insect and possum predation.

The first small octagon was constructed beneath one of the few old midpaddock Red Gums still alive on 'Helm View', to encourage and protect natural regeneration. It is interesting to note that initially the octagon was constructed of 'Ringlock', the only non-electric fence on 'Helm View', but cattle jumped in and destroyed a promising strike of direct-seeded seedlings as soon as the difference between the amount of feed in the clump and that in the paddock was significant. The Milnes responded, with the assistance of Bob Piesse, by using a small solar collector panel to charge a 6-volt battery which electrified the fence, keeping it secure from then on.

The design of this clump is depicted in the diagram below, although with the electric fencing employed at 'Helm View', bedlogs were not required.

The clump can be fenced with high tensile 2.5mm plain wire in a spiral (droppers may be needed in some soils), or with fabricated fence. Wires should be free to move around the posts, not stapled tight to each post—a hard plastic sleeve between the wire and the post allows free movement. Wires should be kept tight with in-line spring tensioners.

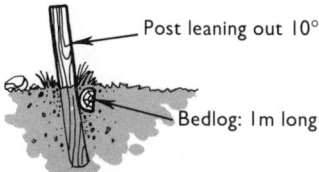

Octagon
Plan view

50m circumference
16m diameter

6–25 m

Post leaning out 10°

Bedlog: 1m long

If possible both posts and bedlogs should be half-round with the flat side facing inwards.

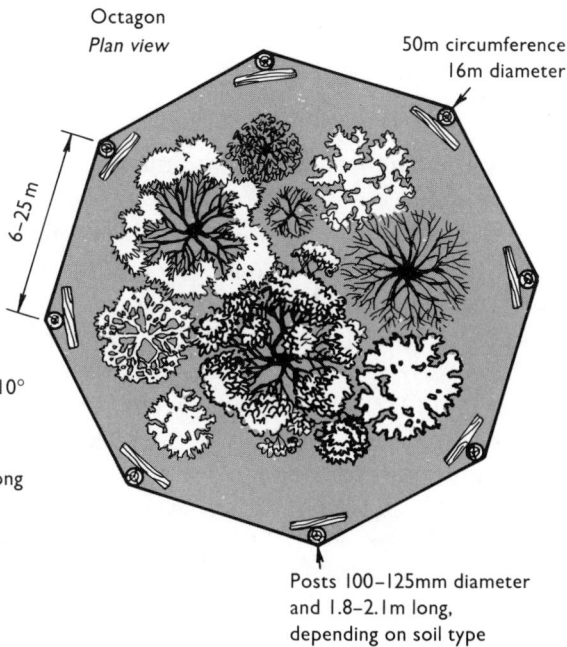

Posts 100–125mm diameter and 1.8–2.1m long, depending on soil type

• A small octagonal clump

SITE PREPARATION

Site preparation and establishment techniques were probably the two most important factors influencing *initial* survival rates and growth rates of trees on the demonstration farms. Careful attention to these almost eliminated the need for maintenance and replanting. Good site preparation enabled planting or seeding to be carried out faster and usually to better effect. Conversely, careless planting techniques on one or two occasions rendered the best site preparation irrelevant.

The site preparation for tube-stock planting was fairly uniform during the first three years of the Potter Farmland Plan. The standard recipe for preparing a site for a planted shelter belt or wood lot was as follows:

- *Autumn:* Rip along planting lines with a single-tine ripper as deep as possible (at least 400 millimetres), to ensure a good shattering of subsoil and subsequent root and moisture penetration.
- *Late winter:* Heavily graze the area to be planted, or spray along the riplines with a knockdown spray such as glyphosate at a rate of 2 litres per hectare.
- *Early spring:* Spray along the riplines with a second application of glyphosate or a combination of amatrol and atrazine at 6 litres and 3 litres per hectare respectively.
- *Mid-late spring:* Plant, and guard if necessary. If conditions are very cold and wet, wait for warmth.

The combination of a deep ripline and split application of herbicide seemed to provide a very favourable environment for young seedlings; free of weeds over the first spring and summer, with good moisture retention in the subsoil beneath the roots and the incentive and capability for roots to seek that moisture. It was noticeable on the demonstration farms that early vigour could be equated with good ground preparation. More particularly, poor early growth rates could often be related to suboptimal site preparation. In the few instances where we did not rip, growth rates were lower, as they were in areas where the weed control was inadequate.

Deep ripping in the autumn before planting, when a good shattering of the subsoil allowed winter rains to soak well down, provided a moisture bank on which seedling roots could draw in the summer and autumn after planting.

Weed control in the one square metre occupied by the tree was critical to the amount of moisture available to young seedlings. If the soil immediately around the tree is *completely weed-free*, then it can store a large amount of moisture, all of which is available to the tree. Work in dry areas of South Australia has shown that on sites which are completely free of weeds, planting or seeding can take place in spring rather than autumn, which is the traditional planting time, *without* watering during the first summer, provided trees are watered in well. The Potter Farmland Plan farmers certainly experienced the benefits of good weed control in several instances where planting took place in December or January without follow-up watering.

Usually a combination of a 'knockdown' or non-residual herbicide and a residual herbicide was most effective, although care was taken that seedling roots did not come into contact with soil contaminated with residual herbicides. On the demonstration farms the most effective combinations were glyphosate and simazine, and amatrol and simazine. Residual herbicides were avoided on poorly drained soils after a trial at 'Willandra', in which one section of ripline was sprayed with a high rate of a commercial amatrol/ atrazine blend. Several inches of rain fell on the poorly drained area immediately after planting, which incorporated the herbicide into the topsoil before seedlings had a chance to get their roots down, and significant losses occurred. We found that the most effective way to combine trees and residual herbicides was to use the 'Hamilton Tree Planter' (see page 115), which was developed for use with residual herbicides.

ESTABLISHMENT

The ground-preparation techniques used on the Potter Farmland Plan farms were not new; they were tried and true methods used by successful tree growers for many years. However the methods of getting the trees into the ground have been refined almost to an art form by John Marriott.

With a diverse range of sites scattered over a large area, many different species, limited time and skilled labour and a high turnover of unskilled volunteers on weekends, it was critical that the project officers develop an almost foolproof system for planting trees successfully.

The Australian Trust for Conservation Volunteers (ATCV) was employed extensively in May, August and September 1985. ATCV crews made up predominantly of 14 to 17 year olds on school holidays camped in empty cottages at 'Nareeb Nareeb', 'Ballantrae', Haeuslers' (next door to 'Willandra') and in shearers' quarters at 'Satimer' and 'Melville Forest', owned by the Tully family. The cost to the Potter Farmland Plan, as to other ATCV clients, was $10 per person per day, plus $0.20 per kilometre for the bus. As expected, the enthusiasm, capacities, work rate and quality of work of individuals varied, as did the collective capabilities of the crews. Whilst it was extremely satisfying to work with young volunteers, it was also very demanding on project staff as a lot of supervision was required. The ATCV was also involved in the Commonwealth Community Employment Programme during 1985 and

1986. The crew members funded by that programme were not volunteers, in that they received a full wage, with a living-away-from-home allowance. Many of these people had been unemployed for some time and they required intensive supervision.

The ATCV is an extremely worthy organisation with a wonderful philosophy. To be fair to the ATCV it should not be regarded as a cheap labour force and it is more suited to smaller projects with less severe time constraints than the Potter Farmland Plan was operating under between 1985 and 1987. Since the demise of the Community Employment Programme, the ATCV has gone back to a reliance on volunteers, and in 1988 achieved a total of 7300 volunteer days' work, planting 270 000 trees, among other tasks. The ATCV offers opportunities for young people to get personally involved in practical conservation projects in all states, which is a tremendous contribution to land conservation.

It became clear after 1985 that additional labour for tree establishment would have to be found. Initially it seemed that mechanical planting would be the answer, as it was claimed that several machines on the market would plant trees reliably at a rate of more than 500 per hour. After a preliminary examination of the different machines available, we purchased a machine from Western Australia on a trial basis, at a cost of $1800 if it proved to be satisfactory. Its initial trials in 1985 showed promise, but it could not operate effectively in the heavier soils and perennial pasture sward of western Victoria, compared with the sand and annual grasses of Western Australia. We decided to modify the machine rather than return it, as local investigations revealed that to build a machine to our specifications would be twice as expensive. Jim Hoffmann, a local bush engineer/inventor, removed the weed-clearing blade and cultivating tine, constructed a much heavier, rigid, planting boot, added a 300-mm coulter, and made the rear press wheels more adjustable.

The modified machine was used extensively in 1986 in a wide range of soil types and moisture conditions. It was a vastly improved machine, capable of getting trees into the ground quickly with a minimum of labour. For example, John Marriott and Peter Waldron planted well over 2000 seedlings by themselves in two days, in very difficult, wet conditions. But the machine was still only planting trees well in a very narrow band of soil types and moisture conditions. Any variations from optimum conditions gave the machine difficulties. Almost every seedling had to be firmed manually.

Another disadvantage of the machine was its lack of manoeuvrability in small shelter belts, especially if the fencing was already complete. Planting machines are suited to large blocks of trees of one or two species, rather than short linear designs with six to ten species. Turning a tractor around inside a shelter belt in wet conditions with deep riplines 3 metres apart is a treacherous operation and getting bogged is always a distinct possibility. For 'fiddly' jobs such as small shelter belts, the speed advantage of a machine over hand planting is not so significant.

The most important drawback with the machine was that it simply did not plant the trees well enough. This initial intuitive judgement was backed up by subsequent observation that machine-planted trees grew more slowly than hand-planted ones. It was particularly striking at 'Willandra', where a large number of machine-planted 1986 seedlings were quickly overtaken by hand-planted 1987 seedlings, despite similar establishment conditions.

Experience with planting machines in other areas is much more favourable. The Chatfield planter from Tammin in Western Australia, for example, has proved to be extremely efficient and successful in the more friable, sandy soils of the wheatbelt. Chatfield machines have planted hundreds of thousands of trees with very high success rates and many Land Conservation District Committees in Western Australia are now purchasing planting machines to hire out to land users.

It was clear during the 1986 plantings that with an accelerated programme, the combined efforts of ATCV crews and the planting machine would not be sufficient to plant the number of trees required in the time available, so community groups were conscripted. Tree planting reunions for contemporary foresters and their friends have been an annual event on my family farm since 1983, and Stuart Cuming had the help of the Glenthompson Football Club in planting trees at 'Fernleigh'. A trial run involving the local community directly in carrying out the revegetation programme began in late spring 1986, when several local sporting and service clubs were engaged to plant trees on the demonstration farms, for $0.40 per tree, including guarding. Their reaction was very positive and it was probably the most effective local extension programme the Potter Farmland Plan has run, as people seemed to learn much more through direct experience than through talking to others, looking over the fence or reading pamphlets. Most of the people involved were locals who asked questions about the reasons behind certain operations, and the basis of the project, as well as the planning process and the technicalities of planting trees. There was a large proportion of farmers involved in a spirit of friendly competition and the output per person was very high. Another advantage was that the project was seen to be spending money locally, as the Potter farmers paid groups directly (fifteen cents per tree to plant, twenty-five cents per tree to guard).

The reliance on inexperienced tree planters from diverse groups was feasible because of a hand-planting system which had evolved in the Hamilton area since 1984, and which was refined into a simple integrated system by John Marriott, the Potter farmers and many groups of volunteers. It is now a tried and proven system which has been used by a wide range of people in a wide range of conditions, planting trees very successfully at an average rate of over fifty trees per person-hour.

The system was designed for people working in pairs with one small wheelbarrow full of trees and one Hamilton Tree Planter. The wheelbarrows which we found to be ideal were square and box-shaped, with flat bottoms and vertical sides, with two wheels and one handle, rather than the more common design which has one wheel and two handles,. We bought them from a Hamilton hardware store for about $55 each. The advantage of this type of barrow was that it was much more stable on slopes, it could be handled with one hand rather than two and its shape lent itself to carrying trays or boxes of trees. Two to three boxes of trees were placed in the barrow, which in dry conditions was filled with 2–3 inches of water to ensure that the root mass of the seedling was very moist, which is essential. A rope was tied to the wheelbarrow and trailed behind as the pair moved along the planting line. The rope was the same length as the desired distance between trees, with a short length of chain attached to its trailing end to keep it taut, and it was used to ensure even spacing between trees where that was desired.

When the first tree in the row was planted, the person with the barrow pulled it along to a point where the end of the trailing rope was level with the last tree, selected a seedling, and removed the tube while the person with the Hamilton Tree Planter selected the best micro-site along the ripline for the tree, and made a hole for the seedling. The person with the barrow handed the seedling to the planter, who planted it, firmed the ground around it and created a small depression for water retention while the barrow person moved along to the next site.

Two keen people with minimal experience could move along planting trees almost at walking pace and achieve very high survival rates, due to the integration of several elements in the planting system. The most important prerequisite was thorough ground preparation, as outlined above, which ensured a weed-free zone along riplines which had had sufficient time to settle down and absorb a bank of moisture in the subsoil. In well-prepared soil, the Hamilton Tree Planter almost eliminated soil disturbance and consequent air pockets, and thus ensured that the seedling obtained maximum benefit from available moisture. The design of the Hamilton Tree Planter is such that the easiest way to plant a tree is the best way—there are no short cuts which a shoddy operator could take at the long-term expense of the tree.

The Hamilton Tree Planter was developed by Keith Cumming, a research scientist at the DARA Pastoral Research Institute at Hamilton. Keith and his colleague, Dr Rod Bird, have been actively engaged in farm shelter, agroforestry and direct seeding research since 1982, which has involved them in a large tree-establishment programme. Keith and Rod have also conducted trials of residual herbicides for weed control and this led them to look for a means of planting trees in which herbicide could be applied before planting and which minimised soil disturbance, avoiding the possibility of contact between seedling roots and the soil containing residual herbicide. Other hand-planting tools such as 'dibble sticks' create a hole for the seedling by compressing the soil, but Keith developed prototypes which removed a core of soil to create a site free of residual herbicide. These initial prototypes worked sufficiently well to encourage Keith to develop models for a range of seedling tube sizes, which began to be used by keen tree growers in the Hamilton district with great success.

People in the Potter Farmland Plan and others in the Hamilton district were using many Hamilton Tree Planters during 1985, and the growing demand led Keith to establish a private company to manufacture and distribute the planters through hardware stores and farm merchandisers in Victoria and south-eastern South Australia.

In dry conditions, planting with the College Cricket Club at 'Barnaby', a watering container was placed in the wheelbarrow with the trees, and used to fill the hole made by the Hamilton Tree Planter before the tree was planted. This was an extremely efficient way of watering the seedling in, as most of the moisture soaked down below seedling roots where it was most useful. On a couple of dry sites this system was used to great effect and 300–400 millilitres of water was a sufficient quantity to get young trees off to a very good start in difficult conditions. 'Watering-in' is essential in dry areas and is the best insurance for young seedlings when carried out well. The most efficient and effective way to water trees in is to 'puddle' them, putting the water in the planting hole *before* the seedling, ensuring that the tree is placed into a mud pie, ensuring plenty of moisture around the roots where it is needed most and minimising air pockets and evaporation losses. It may be slightly faster to water trees after planting, but most of the water poured on to the soil around the seedling stays near the surface, rather than saturating the root zone where it is much more effective and less prone to evaporation.

The intensive revegetation experience of the Potter Farmland Plan demonstrated that with good planning and some successful experience to look back on, planting trees on farms need not be the chore that it is often considered. We found that with this hand-planting system, the labour involved in tree establishment can be provided by all members of the family and friends, and several hundred trees can be planted in a day very easily. Some Potter Farmland Plan farmers had annual tree planting programmes of several hundred trees before their involvement in the project, but now say they can comfortably establish several thousand trees per year without external support.

The individual components of the hand-planting system developed by John Marriott and volunteer crews over the first three years of the project all existed prior to 1985. Keith Cumming had invented and refined the Hamilton Tree Planter in 1983 and 1984, and the compatibility of the planter with residual herbicides had already been proven in Pastoral Research Institute trials. Our experience with groups of ten to twelve people at a time, many without practical knowhow, was the anvil on which the Potter Farmland Plan system was forged. John Marriott was attracted to the small wheelbarrows by their shape and price. We had decided very early in the project that we would use the Hamilton Tree Planter with residual herbicides and that a 'Micron Herbi' would be the best means of application. The use of a rope behind the barrow to determine tree spacings was an innovation of John's, when we found several times that we ran out of trees well before the end of the ripline, because crews preferred to plant trees close together so that they could chat. It was a logical extension to modify the wheelbarrows by welding a support on to each side, upon which a stick or rod could be placed, to create a 'toilet-roll holder' arrangement from which the rolls of tree guards could be easily attached, just like the plastic bags in the vegetable sections of supermarkets.

The development of a hand-planting system which can be used by groups of unskilled people successfully planting from fifty to one hundred trees per person per hour for a total material cost of less than $150 is a major achievement. It is a testimony to John Marriott's patience and persistence over many wet, Western District weekends and his continual striving for a more efficient way.

Direct Seeding and Natural Regeneration

Direct seeding is the establishment of vegetation by sowing seed directly onto the site. The key difference between direct seeding and natural re-generation is that in the latter, the seed comes from vegetation on or near the site, and natural seed distributors—wind, insects, birds and small animals—are relied upon to do the sowing. In direct seeding, the seed is usually collected off-site, and is brought to the revegetation site and sown by humans. The methods of seed placement include simply broadcasting by hand ('feeding the chooks'), broadcasting through farm machinery, drilling in rows with machinery designed specifically for the purpose and precision placement in niches (often with seed mixed in a medium designed to minimise weed competition, moisture loss, seed predation and wind removal). In most cases, in both direct seeding and natural regeneration, particularly on farmland, the seedbed is prepared with some form of soil disturbance and/or herbicide to control weed competition.

Thus direct seeding and natural regeneration range from simply removing browsing animals from around old trees to sophisticated techniques involving specialised machinery, seed treatment, vermin and weed control and precision placement of seed selected for particular traits such as salt tolerance.

There is nothing new about direct seeding or natural regeneration. Natural regeneration has been occurring in Australia for millennia. However in many areas, domestic and feral animals have continuously grazed emerging seedlings and removed understorey vegetation, and introduced crops and pastures have out-competed native species. Direct seeding has been used for tree establishment in Australia for most of this century. Broadcasting seed on ploughed ground was a common method of establishing large shelter belts of *Eucalyptus cladocalyx* and *Eucalyptus globulus* in South Australia and south-west Victoria early this century. Forestry organisations have been direct seeding to regenerate logging sites for many years.

The introduction of superphosphate and subterranean clovers in the 1930s put an end to direct seeding on farms, as competition from weeds became a limiting factor. It was not until the late 1970s that a few innovative farmers began to experiment with tree seeding again, encouraged by the potential of the new herbicides coming on to the market and frustrated by the time, labour and money involved in planting individual seedlings. Mining companies also turned to direct seeding in the early 1980s to rehabilitate mining sites and some statutory authorities began to use direct seeding to stabilise steep slopes such as road batters.

However, the acceptance of these methods as practical alternatives to planting trees has been slow. In many areas of Australia the dominant ethic still leads to the destruction of native vegetation rather than its preservation or re-establishment and in the areas where revegetation is occurring, planting seedlings either by machine or hand is still by far the most common method. It is clear that most people either do not know about direct seeding and natural regeneration, do not believe that these methods will work, or do not have the confidence, skill and resources to have a go.

When the Potter Farmland Plan began in 1984, there were a few examples of direct seeding on Western District farms. Bill Sharp, in his role as secretary of the Glenelg Farm Tree Group, had established a number of direct-seeding trials on Hamilton district properties, and Richard Weatherly was experimenting with different weed control techniques near Mortlake. Some farmers had accidentally achieved spectacular natural regeneration, usually of Red Gum seedlings on soil prepared for crop or pasture establishment.

So we had some information and we were keen to have a go. In 1985 we direct seeded 31 hectares on the demonstration farms, using a range of methods, including the Western Tree Seeder (a single-row seeding machine developed by a syndicate of Victorian farmers including Richard and Bill Weatherly, Bill Speirs and Bill Sharp), hand broadcasting, and farm combines, with various combinations of mechanical and chemical weed control. This was a significant proportion of our revegetation effort and, as we only had three years of Potter funding in which to achieve a result, our hearts were in our mouths hoping for success.

Initial results were not encouraging. Assessment crews found very patchy clumps of small seedlings amongst vigorous weeds and whilst we could see the undoubted potential advantages of the technique, the farmers, John Marriott and I were all disappointed. We could not afford three years of failed direct seeding on the demonstration farms, so we reduced our emphasis on direct seeding and natural regeneration, seeding ten hectares in 1986 and six hectares in 1987.

In a couple of instances the 1985 results were so disappointing that we decided to spray the sites with glyphosate and start again. Bruce Milne did this in spring 1987 at 'Helm View' and to his surprise found some young eucalypt, casuarina and acacia seedlings left standing among the herbicide-killed weeds. This experience was repeated several times, so we began to re-assess the effectiveness of our early direct seeding. Some of the sites thought to be failures now look quite satisfactory five years later. The more random pattern of germination achieved with direct seeding and natural regeneration compared with regimented tube-stock plantings usually leaves a few gaps and a few impenetrable thickets, which is great, unless a shelter belt of constant permeability is required.

While we were having mixed results with direct seeding, others were developing more reliable techniques. Peter Waldron on 'Willandra' is convinced that it was the only way to go in the long term and he persevered on a small scale, reverting to the use of a mouldboard plough to completely invert clods of soil, burying weed seed and providing a range of niches for germinating seedlings. An advantage of this method is that it avoids the use of herbicides, but it does require a skilled operator to achieve good weed control with a mouldboard plough. Peter achieved great success with this method and has hardly *planted* a tree since!

Dr Rod Bird and Keith Cumming of the Pastural Research Institute at Hamilton were also engaged in comprehensive direct-seeding trials, referred to in Chapter 6. Since 1987 they have tested a range of direct-seeding methods over many sites and their most successful treatments (including mouldboard ploughing) have been extremely reliable. The direct-seeding recipe distilled from their results is most suitable for south-western Victoria, but the elements of successful weed control are more widely applicable.

A rough guide to successful direct seeding in south-western Victoria from results to date is as follows. Initial site treatment should be mouldboard ploughing in August-September (the choice where the land user dislikes herbicides or the soil remains inundated for extended periods), or deep ripping in autumn, heavy grazing in late winter then two applications of glyphosate at two to four litres per hectare in 1.5 metre bands along the riplines, with the second application including a residual herbicide such as simazine at seven litres per hectare or atrazine at five litres per hectare. To sow the seed in mid-September to mid-October, scalp away a narrow (say 10-centimetre) strip of soil containing weed seeds and residual herbicide and roughen the surface using tines behind the scalping disc. Mix the seed with equal parts of sand or sawdust for sowing and firm it in with a press-wheel or a bag containing a little sand. If weeds look like becoming a problem, the site can be oversprayed with simazine *before* autumn germination of weeds.

If broad-scale sowing is preferred to lines, a combine or scarifier can be used to roughen the seedbed in September, sowing tree seeds by hand or via a small seed box. Weed problems can be reduced by a light application of glyphosate in the spring before sowing to reduce seed carryover, and if feasible, an overspray of simazine in the early autumn following seeding will give an extra year of critical weed control.

Much more detailed information on weed control, seeding rates, seed viability and techniques for low-rainfall sites can be obtained directly from Rod and Keith. Their work has shown that direct seeding of trees can be as reliable as tube-stock planting if the same care and attention is taken with site preparation and weed control as good farmers take when sowing a crop.

Protection from Grazing Animals

The Potter Farmland Plan farmers constructed more than 164 kilometres of fencing between 1985 and 1987, more than half of which was to protect 63.8 kilometres of shelter belts, 37 hectares of clumps, 47 hectares of direct seeding and 9 hectares of wood lots. This intensive experience in a range of conditions led to a rigorous examination of the design and construction of reliable tree protection systems and a recognition of the need for significant improvement in this very expensive component of farm tree establishment. This section outlines the methods used on the Potter farms and the knowledge that was gained along the way.

Individual trees

The PFP constructed 2182 individual tree guards from 1985 to 1987. Our farming districts are dotted with hundreds of different types of guards for single trees, from corrugated iron tanks, to old oil drums, tractor and truck tyres, 'Weldmesh', loops of electric wire, posts and rails, logs, rocks and even disused Bainbridge wheels from irrigation channels. Some are cheaper than others, some are more attractive than others and some are more effective than others. There is no 'right' guard except one which does the job, which is to protect the tree for as long as is necessary. The design of a cheap, effective stock-proof tree guard is a challenge which is likely to be met in many different ways in different areas as protection needs and available materials vary. A book published in 1988 by the ABC and Greening Australia, *Caring for Young Trees*, contains many innovative examples of tree guards which have worked in a wide range of environments.

Common mistakes in the design and management of single tree guards include:

1   Designing a guard which has to be removed before the tree is big enough to fend for itself; 200-litre drums are a good example, as many people remove them while it is still possible to lift them off, which is often too early for the protection of the tree. People then have to provide the tree with additional protection, which suddenly makes the originally cheap guard expensive. The alternatives are to leave the drum on permanently, to cut the drum in half before using it to enable easy removal, or to cut it from the tree, all of which add expense.

2   Removing guards too soon. Many people purchase or make guards with the intention of rotating them around the farm every three to five years. This is a good idea in principle but it is often tempting to remove guards too early. Most trees are vulnerable to stock damage from rubbing (particularly by bulls and rams) even when the trees are 150 millimetres (6 inches) in diameter at rump height.

Some species such as stringybarks are vulnerable to bark damage irrespective of how big or old they are. Wire netting is sometimes used to protect bark, but it tends to strangle trees when left on too long.

• Hand-planting trees is simple, cheap and effective with good planning, the right species, good weed control and appropriate tools. Jackie and Angus Brown of Hamilton are an efficient planting team with the Hamilton Tree Planter, the two-wheeled barrow and the spacing rope shown below.

• Four more aspects of direct seeding: seeding single rows with the Western Tree Seeder; using a conventional farm sod-seeder to broadcast seed; John Marriott broadcasting by hand with a lawn fertiliser spreader; and current direct seeding research trials carried out by the Pastoral Research Institute at Hamilton.

DIRECT SEEDING OF TREES

DEPARTMENT OF AGRICULTURE & RURAL AFFAIRS
NATIONAL SOIL CONSERVATION PROJECT

RESEARCH
A. Site preparation, seedbed & sowing factors
B. Spring - Summer weed control using residual herbicides before sowing
C. Autumn -Winter weed control using overspray herbicides

SPECIES
36 species of trees & shrubs from Acacia, Bursaria, Casuarina, Callistemon, Eucalyptus, Leptospermum & Melaleuca

SITES
Southern Wimmera - Nurrabiel   Basalt plains - Buckley Swamp
Dundas tablelands - Melville Forest   Ordovician sediments - Glenthompson

• Peter Waldron's old-fashioned mouldboard plough has found a new lease of life preparing soil for direct seeding. The photograph above, taken in November 1986, shows the soil just before sowing. The shot below shows the results on the same site in winter 1989. There is a good mix of eucalypts, acacias and casuarinas in this belt.

• Community involvement in revegetation pays off for farmers, the landscape and for general awareness of the environment. The Glenthompson Football Club stalwarts on 'Reedy Creek' (above) and the Dunkeld School at 'Cherrymount' (below) were among more than twenty organisations which helped to plant the demonstration farms.

Old trees can be protected by placing old posts vertically against the trunk as a barrier against bark damage from rubbing. With the growth of interest in farm trees, it is likely that some trunk protection products could be developed and marketed for use on farms—for example stiff plastic, pre-stressed into a spiral cylinder which can expand as the tree grows (possibly impregnated with animal repellents?). Peter Waldron constructed an effective trunk protection sleeve by cutting a slit length-wise along 100-millimetre PVC stormwater piping. Peter used this technique on 'Willandra' Red Gums after 200-litre drum guards were removed.

3   Making guards too small. Guards should provide enough protection for the tree to develop a big enough crown with which to grow. It sounds simple, but on many farms, for the sake of saving 40 or 50 cents per tree, the guards used are too small and stock are able to browse at least some of the leaves, which reduces the 'growing power' of the tree. Guards should be either repellent to stock, higher than the animals can reach when standing on their hind legs, or wide and strong enough to allow for the full length of the animal's neck and head to lean over the guard without reaching leaves.

4   If mesh guards are used, having too large a mesh grid. We found on the demonstration farms that 100-millimetre grid squares were most effective, as sheep could get their head stuck in 150-millimetre squares and were easily able to browse trees through 200-millimetre squares. Thus, although 100-millimetre mesh (such as 'F41') was more expensive, larger squares were a false economy for sheep and goats (but not for cattle or horses).

5   Making guards too flimsy, or not fixing them sufficiently firmly to the ground. Much of the pressure on single-tree guards is due to stock rub-bing. When bulls, rams or horses are attracted to tree guards for rubbing or testing their strength, guards either have to be very robust indeed (which usually means expensive), or repellent to stock. Good manage-ment—particularly lice and tick control, should minimise rubbing. Some farmers nail bottle tops to dead trees or stumps to provide better alter-native sites for rubbing, thus easing pressure on tree guards. Galvanised nails set in lines are more durable than bottle tops and have the added advantage of allowing clearance of hairs by birds for use in their nests.

In an interesting sidelight, Bill Speirs won a Nuffield Farming Scholarship (awarded to two Australian farmers every two years) in 1988 to study farm planning and land conservation in Great Britain. He noted tree guards made of a translucent tube of corrugated plastic, which come in a range of diameters and heights, marketed under the name of 'Tubex'. Bill was so impressed with these guards that he became the Australian agent for them and has been importing them by the container load to satisfy a rapidly increasing demand.

Clumps

The advantages of clumps over individual trees have already been discussed, and more than 100 mid-paddock clumps were established on the demonstration farms from 1985 to 1987. We found that the problem of fencing small clumps was to keep the fence tight, despite short strains, in a cost-effective manner. Circular clumps are efficient in that each post takes an equal portion of the strain, thus eliminating expensive strainers and stay assemblies. They need not be perfect circles, but can be any rounded shape. The Potter Farmland Plan experience with circular clumps was initiated after experimentation with oblong and triangular natural-regeneration clumps on 'Pine Grove' in 1985. The octagonal design, 50 metres in circumference, which we discussed earlier, was first constructed on 'Helm View' in spring 1985.

Circular clumps could be much larger than the one shown on page 109, as long as the proportions remain the same. On the demonstration farms where sheep were the only stock the bedlogs illustrated were not necessary, but in plastic or sandy soils the bedlogs would be essential with any stock, unless the fence is electrified. At 'Helm View', electric fencing enabled clumps to be established with 1.8-metre posts and no bedlogs, for both sheep and cattle. Stock often put intense pressure on small clumps, since grass grows well inside the clump and is very attractive if the rest of the paddock is bare. Consequently the rules which apply to all tree fences are even more critical for small clumps. Wires, particularly bottom wires, should not slacken. They can be kept tight with the use of mid-line tensioners (heavy-duty springs are the best). Experience on the demonstration farms has shown that using rounded shapes in this way is still quicker than erecting strainer assemblies and it allows for innovative designs sympathetic to the lie of the land.

Another approach has been adopted by farmers in the Whitehead's Creek catchment near Seymour, Victoria, who have established clumps of similar shape and dimension to the Potter Farmland Plan design, but with steel 'star pickets' leaning inwards at 45 degrees, supporting five or six hand-tight strands of barbed wire. One farmer has had more than fifty such clumps in place with sheep and cattle for more than four years without stock penetrating any clumps.

Fences do not have to be curved, however, and strategic mid-paddock shelter can also be created effectively with more conventional fencing.

Electric options

Electric fencing is ideally suited to the protection of small clumps of trees. One of the most important advantages of electric fencing for protection of clumps is the capability to fence around curves and in odd shapes without the added expense of end assemblies. Design A on page 106 at 'Helm View' was protected by a four-wire electric fence. It has successfully protected trees from sheep and cattle for two years, even when extreme pressure was placed on the fence during periods of 'crash grazing' at very high stocking rates to prepare pasture for direct-drilling of new pasture species. The posts were 1.8 metres by 100–125 millimetres, driven to a depth of 1.1 metres, spaced at an average interval of about 7 metres. The posts were closer together on the sharpest points of the curve. This clump was powered by a lead-out wire (with an isolator switch) from a laneway about 60 metres away, but it could easily have been powered by a small solar energiser.

Protection from Vermin

The word 'vermin' in this section refers to introduced animals other than stock that browse trees, such as hares and rabbits; and to native animals such as kangaroos, wallabies, possums, cockatoos and magpies which also damage farm trees. The main hazards on the demonstration farms were hares, cockatoos and magpies.

The first step in protecting trees from browsing by vermin was to determine the hazard. Where the farmers considered that vermin were a hazard, the next step was to decide whether to protect individual trees with guards, or to protect whole blocks of trees by altering fence design to exclude vermin. Using the project's costing for tree guards, the only type of planting for which vermin-proof fences were more cost-effective than guards was in square or circular blocks. Barry Levinson and David McDonald decided to plant without guards, concluding that it was easier to plant more trees and allow for losses than to construct a guard for each tree.

The need for protection of trees was minimised by careful attention to plant establishment techniques. The complete removal of weeds from the area to be planted before planting leaves an obvious area of bare ground which tends to act like a magnet for birds looking for worms and grubs, and leaves the newly planted seedlings standing on their own, almost asking to be nipped off by hares, rabbits, cockatoos and corellas. The most critical area for weed control is a one-metre circle around each seedling. We decided that it was better to spray or cultivate in strips along riplines or in spots, rather than spraying or cultivating the entire area. We therefore used a

'Micron Herbi' to spray in strips along riplines, allowing the weeds between the rows to grow freely.

This selective weeding reduced the vulnerability of the trees to vermin, since rabbits and hares will not fight through thick, long, wet grass or cape-weed to get to trees, and birds are less likely to be attracted to the site looking for grubs. Trees were also less vulnerable to the effects of wind—the weeds served to shelter young seedlings.

Techniques used on the Potter Farmland Plan farms for protection against different types of vermin were as follows.

Rabbits and hares

Satisfactory guards include clear ultraviolet-treated plastic sleeves supported by three stakes, wire netting with a wire stake and old car tyres, all of which are re-usable. We used the plastic sleeves almost exclusively, many of them three times. They were very effective and several of the farmers say that they would no longer consider planting a tree without such a guard. Cheaper guards (such as old fertiliser bags and 'bottle-top' guards) cannot be re-used and are not as reliable as the others mentioned. Old fertiliser bags perish quickly, and may not provide protection for long enough, and the bottle-top guards rust quickly, are hard to remove, and often ring-bark young trees as branches grow out through the holes. Car tyres are effective, but should be cut through before use, to enable easy removal. Care should be taken when removing them, too, since snakes like to camp in them, which can make removal an adventurous operation!

It was very easy to remove the plastic sleeve from three stakes, but often difficult to extract stakes from the ground without getting splinters and/or a sore back. To alleviate this problem, Andrew Milne conceived, designed and constructed some metal clamps attached to a chain and handle, which, when placed over a stake, allowed for easy removal, applying increasing grip to the stake as the pull strengthened. For very tight stakes, Andrew developed a lever and fulcrum design, which gripped the stake at its base, and lifted it out of the ground as the operator pressed down on the handle, in the same way as a see-saw goes up at one end when someone sits on the other end.

Initially we used hardwood stakes, some cut from the Monier mill at Hamilton, others from diverse sources—for example we were able to obtain 20 000 reject broom handles (made of kiln-dried and finished hardwood) for 10 cents each from a factory in Melbourne. These hardwood stakes could often be cut in half and were much more cost-effective when used several times. As the project developed, the local supply of stakes dried up and project staff began using bamboo stakes of 8–12 millimetres by 900 millimetres, obtained for 7 to 10 cents each. The bamboo stakes were rarely re-usable, and removal was easy; however pushing them into the ground can be difficult and can lead to bruised hands (after several hundred stakes) or deep splinters if the stake breaks with someone pushing down hard on it. With the safety aspect in mind, we designed a T-shaped device made of 19-millimetre piping to fit over the top of the stake, so that we could push it in with both hands, applying even pressure from each side, eliminating the possibility of fracture or splintering and generally doing a better job. These 'T pushers' were manufactured locally for the project.

Another problem in using crews of volunteers of varying experience to place guards around trees was the apparent inability of many people to judge the first two corners of an equilateral triangle, when banging or pushing in the first two stakes. This meant that many guards were quite long and narrow isosceles triangles, often with the seedling in one corner, allowing it to flap against the plastic, and possibly to fry if pushed against the plastic in very hot weather. To solve this problem, John Marriott designed a metal equilateral triangle of the correct dimensions, which was placed over the seedling before putting in the first two stakes, to show where to place the stakes. These triangles became part of our planting system, ensuring that trees were not only planted well, but also protected by sound guards.

While the project developed and demonstrated very efficient methods of applying existing techniques to the protection of trees, project staff believe that there is plenty of room for improvement in tree-protection systems and more innovative, cost-effective techniques are needed.

Rabbits and hares were deterred at 'Daryn Rise' by blood and bone buried near the newly planted seedling (about a matchbox full, buried 2–3 centimetres deep, 15 centimetres from the tree). This repellent works for only two or three months, but this seemed to be enough to get trees through the period of maximum hazard.

Birds

The most effective technique for preventing bird damage has already been mentioned. This was to avoid attracting 'nuisance' birds such as cockatoos, corellas and magpies to the tree areas. Plastic guards prevented damage to very small seedlings, but damage occurred at 'Daryn Rise' when trees were well out of the guard.

There is great potential for research into specific repellents which work by smell, taste or sound for different species of vermin. Such research is under way already in many countries; for example, pine oil is being investigated as a deterrent to deer and hares in conifer plantations in British Columbia and paint and egg-based formulas have been used successfully in New Zealand plantations. It is likely that more efficient and effective vermin repellent systems than the present physical barriers will emerge over the next ten years.

Protection from Climate

The first step in protecting trees from the climate is planning, which was done well before the trees were established on the demonstration farms. Species selection, location of planting/seeding sites, timing and establishment technique were the most important factors to be taken into account. Western Victoria is a favourable environment for farm-tree establishment, the main climatic hazards being waterlogging and frosts in winter and strong winds. These hazards were largely avoided by thorough ground preparation in autumn and by planting indigenous species in late spring. The principles applied on the demonstration farms are applicable elsewhere, and in difficult environments it is even more important that they are followed.

Indigenous species, planted or seeded into similar conditions of soil type, slope, aspect and drainage to those in which they occur naturally, are likely to suffer least from frost, wind and desiccation. Problems can also be minimised by avoiding frost hollows, or very exposed sites, when planning the location of plantations, clumps or single trees, and by planting at a suitable time of year.

Nevertheless, when attempting to re-establish vegetation on farmland, there will be times when young trees have to be protected from frost, wind and desiccation, regardless of siting and species selection. Areas such as the Tasmanian midlands and the tablelands of New South Wales are notoriously frost prone, other regions are arid and in many areas newly established trees are subjected to strong winds. Stress is often due to a combination of these factors.

Frost

The plastic guards around each tree were probably the most effective defence against frost on the demonstration farms and they had the advantage of providing protection against wind and desiccation as well. The most effective plastic guards were those treated against ultraviolet light, which although more expensive than used fertiliser bags or rubbish bags, were re-used at least twice and often three times. Other measures which are used successfully in some areas are old car tyres, 'the grow tube' and bales of rain-damaged hay.

In areas prone to severe frosts we observed the following 'don'ts'. Don't plant frost-tender species and don't plant in frost hollows. There can be a conflict between providing sufficient moisture to seedlings, and minimising the potential for frost damage. Good weed control is essential in some areas to prevent desiccation, but weeds can actually protect trees from frost. We decided which type of stress posed the greatest threat to our trees, and planned accordingly, before considering the 'secondary threat'. For example, from experience in the Hamilton area, project staff knew that frost and waterlogging are more frequent killers of young trees than drought stress. As most frosts occur in early spring, it was preferable to plant in late spring and take the risk of having to water trees to get them through the first summer, rather than planting earlier and risking frost damage.

Wind

Plastic guards provided good early protection from wind and 200-litre drums for single shade trees provided excellent early shelter. We found that it is possible to protect the tree too much, encouraging very fast early crown growth, without the corresponding development of a strong root system. This means that when the seedling grows above the guard it is often unable to cope with sudden exposure to the elements. This occurred with 200-litre drums at 'Gheringap' and 'Willandra', where several young trees were almost ringbarked after emerging from the top of the drum. However, placing a used truck tyre on top of the drum fixed this problem and provided better protection from stock. The smaller 450-millimetre-high plastic sleeves or open mesh guards such as wire netting, open-weave or black polyethylene mesh do not create such problems.

Weeds can also provide wind protection, as well as protection against rabbits and hares, as we have seen. If possible, weed control should be kept to a minimum; treat the area within half a metre of the tree very thoroughly, giving weeds a free reign elsewhere.

'Windthrow' (when trees blow over, usually due to a sudden increase in exposure or during strong winds when soils are soft) sometimes occurs with farm trees when seedlings are planted into deep riplines. This happened on several of the Potter farms, particularly with fast-growing species such as *Acacia mearnsii*. Windthrow can be avoided by planting *beside* the ripline rather than in it, which is preferable in wet areas to avoid waterlogging as well as windthrow.

Desiccation

Minimising soil disturbance by using the Hamilton Tree Planter, as described above, was a key factor in moisture conservation, as it avoided the large air pockets that can be a problem for shovel-planted trees. This meant that a minimum amount of watering was required. In dry conditions, the Hamilton Tree Planter requires as little as half a litre of water in the planting hole to provide sufficient moisture, as all the moisture goes to the base of the seedling root system where it is needed most, rather than on to the soil surface, where it evaporates quickly. Allied with the use of the Hamilton Tree Planter, the other aspects of our system—the deep ripping and thorough weed control, were critical in ensuring a reserve of subsoil moisture, all of which was easily accessible to tree seedlings free of competition.

We tried to improve the water supply to small trees on sloping, dry sites by encouraging planting crews to make small crescent-shaped banks down the slope from the tree to trap run-off where it is useful to the tree. It was possible to make these by hand, because sites were ripped or ploughed before planting.

Mulching is a means of preventing the soil around the base of the seedling from drying out, eliminating weed competition, and thus increasing the amount of moisture available to the tree. Mulches range from woodchips, gravel and screenings to sheets of polythene, old carpet squares, newspaper or hessian or organic mulches such as straw, grass or leaf mould. With at least 650 millimetres average annual rainfall on the Potter farms, we did not need to use moisture-absorbing agents or mulching to conserve moisture; however there are many areas where it is a critical factor.

Good mulching is very effective, particularly on dry sites or where herbicides are not used, but there are some traps.

1   Mulching is time consuming and more than doubles the labour involved in tree planting.
2   Organic mulches such as green lawn clippings should be well-rotted, or they tend to ferment, robbing the tree of nitrogen.
3   Plastic or polythene sheets, including the commercially produced weed mats, provide ideal habitats for marsupial mice and rats. They tunnel under the mulch, often eating tree roots and increasing the tendency to windthrow.
4   In areas prone to grass fires, organic mulches provide fuel which can burn or smoulder for hours close to the tree. This was apparent after

the 1983 Victorian fires, when mulched farm trees were destroyed but others recovered in places where fires had swept through quickly.

5   Organic mulches, if placed against the bark, can cause collar rot.

We did experiment with mulching for weed control at 'Wyola', but there was no significant difference in survival or growth rates, so the results did not justify the significant increase in materials and labour costs.

(5)

Lessons and Rewards

● ON-FARM BENEFITS

SHORT-TERM PRODUCTION

T HE ON-FARM WORKS described in Chapter 4 will have an increasing impact on farm outputs for many years. Some benefits of the whole farm planning approach implemented, however, were immediately obvious.

Renovation of Saline Pastures

The most striking impact of renovating salt-affected pastures occurred in paddocks at Melville Forest which were formerly showing the effects of dry-land salting. The Milnes estimate that the carrying capacity of the first salt-affected paddock, sown in September 1985, more than doubled (to at least ten dry sheep equivalents per hectare) by the end of 1986.

The experience of Peter Waldron is similar. Peter estimates that the carrying capacity of his formerly salt-affected areas has at least doubled, and probably tripled, and he is confident that the lateral spread of these areas has been arrested. However, seeding using similar techniques at 'Wyola', 'Daryn Rise' and 'Reedy Creek' was less successful.

The Potter Farmland Plan has demonstrated the potential for landholders to increase short-term production from degraded land, but success on difficult sites is not guaranteed. It is subject to the same variables that affect most other farm activities. The detailed records of establishment techniques and seeding, herbicide and fertiliser rates compiled by John Marriott have enabled other landholders in the region and staff from the Department of Agriculture and Rural Affairs and Conservation and Environment to learn from the demonstration farms.

Improved Farm Layout

Some of the benefits of improved farm layout have already been discussed, as the renovation of salt-affected pastures would not be possible if they were not fenced into separate management units.

Stuart Cuming, Bill and Jack Speirs, Gavin Lewis and Peter Waldron have all mentioned significant benefits from fencing into smaller paddocks—not only more efficient pasture use but also added flexibility of management, particularly of small mobs of ewes during joining and lambing.

Laneways

Laneways were a major feature of the farm-layout improvements implemented. Twelve of the fifteen demonstration farms already had laneways when the project began and nine of these systems were later expanded. The benefits of lanes are described in detail in Chapter 3. They are particularly useful on grazing properties for a number of reasons:

- Well-designed laneways concentrate stock traffic on those areas which are least prone to erosion, or in areas such as creek crossings which have been modified with rubble and/or pipes to avoid erosion.
- They reduce the time involved in moving stock and the risk of getting different mobs mixed together or 'boxed'.
- They reduce the stress suffered by both man and beast in moving stock. A well-designed laneway system should ensure that each paddock is only one or two gates away from the shed and yards.
- They reduce the time spent opening gates when moving around the farm.

The lanes established on the demonstration farms from 1985 to 1987 began to save landholders time and energy immediately. They were also particularly useful for showing visitors around on demonst   on farms!

Water Supply

Where existing water supplies are subject to sedimentation and salinity, improvements to water supply can immediately lift production. Salt levels in stock water higher than about 5000 parts per million cause weight loss and can lead to weak points in wool fibres, known as 'tender wool', or 'a break in the wool'. The situation on both 'Daryn Rise' and 'Wyola' was not quite so bad as that, however in each case the sole source of water in some paddocks, before the project began, was a saline creek. The whole farm plan in each of these cases incorporated new dams high in the subcatchment and fencing out and revegetation of the saline drainage line.

LONG-TERM PRODUCTION

The benefits of the farm-improvement measures outlined above were immediately obvious, but many of the works implemented from 1985 to 1988 (shelter belts, for example) will have a more gradual impact—their initial effects are subtle but their value becomes more obvious with each year.

Shade and Shelter

One of the major elements of our whole farm plans was the revegetation strategy. Re-establishing ecologically viable stands of indigenous vegetation is very important in most agricultural regions of Australia, but it is particularly valuable in south-western Victoria, given the climate, the type of farming practised and the need to increase plant water use to prevent dry-land salinity.

Farmers perceive shade and shelter as the main reason for planting trees on farms. This has been borne out by a number of studies, the most recent being a study of Howett and Lothian in South Australia in 1988.[1] The Cary, Beel and Hawkins study,[2] a comprehensive study of Hamilton district farmers' attitudes to land management practices, including tree planting, was commissioned by the Potter Farmland Plan in 1986. The table overleaf shows how many farmers in three sample groups cited a range of benefits as their reasons for planting trees.

The table is interesting in that it outlines both the differences in perceptions between the Potter Farmland Plan participants and eighty-two farmers from the Shires of Dundas and Charlton, and the changes in each group between the time they first planted trees and the time of the survey— mid-1986—when the project had been operating for eighteen months.

Farmers' Reasons for Tree Planting (% of mentions)						
	Potter Farmland Plan		Dundas Shire		Charlton Shire	
Reasons for planting trees at first, and reason in 1986	*First* %	*1986* %	*First* %	*1986* %	*First* %	*1986* %
Shade and shelter	44	75	86	68	85	72
Improve the landscape	19	38	27	43	49	33
Replace Trees	25	19	11	16	5	15
Reduce, control salinity	19	38	11	16	3	8
Improve productivity	6	12	3	8	0	0
For the future	6	18	8	11	0	0
Increase land values	0	0	5	5	0	0
Wildlife	0	12	5	16	5	13
Firewood, farm timber	0	6	5	5	3	2
Erosion	25	18	3	0	0	3
Lower water table	6	12	0	0	5	8

Note: The figures refer to the percentage of farmers who mentioned the particular reason for planting trees; most farmers mentioned more than one.

Several points can be gleaned from the table.

1   Shade and shelter were the most commonly cited reasons for tree establishment in all areas. Its relative importance increased significantly for the demonstration farmers and decreased slightly for the other two groups. This suggests that the shelter strategies implemented on the demonstration farms had a strong influence on the farmers' thinking.

2   Landscape improvement was the second most commonly cited reason in all three areas. Its relative importance increased for the Potter Farmland Plan farmers and Dundas farmers but dropped slightly for the Charlton group.

3   The Potter Farmland Plan farmers thought that trees had a wider range of values than did the other two groups, whose perceptions were more focused on shade, shelter and landscape improvement.

4   There was a significant increase in farmers' awareness of the value of trees for wildlife habitat and salinity control.

5   Farmers in grazing enterprises (Potter and Dundas) associate trees with improved productivity and being better off 'for the future' than farmers in cropping districts (Charlton).

It is understandable that western Victorian farmers associate the shade and shelter of trees with increased productivity, since considerable stock losses are caused by cold snaps there every few years. In 1982 severe weather killed many thousands of shorn sheep in the Western District.

On 1 December 1987, official DARA estimates suggested that 30 000 sheep died from cold stress in one day and it was widely acknowledged in the district that the official estimate was extremely low. Farmers in the Cavendish and Balmoral areas, which include the Potter Farmland Plan farms at Melville Forest, lost up to half their total flock in one night. The financial impact of such a loss, even in a good season, is devastating.

The benefits of shelter were graphically illustrated on the property of Neil Lawrance at Gatum, about 10 kilometres north-east of 'Helm View'. Neil was in the middle of shearing approximately 9000 sheep when the storm hit. He forced the shorn sheep into shelter belts, some of which were newly established, and restricted his losses to seventy-four sheep. The young trees in his shelter belts suffered minimal damage, too, because the sheep were too preoccupied with survival to browse. By contrast, both of Neil's neighbours lost many more sheep, in one case nearly 50 per cent of the flock.

The benefits of shelter have been quantified in several studies, one by Dr Rod Bird at the DARA Pastoral Research Institute, Hamilton, and others by New Zealand researchers who have measured 60 per cent increases in pasture production six tree heights away from a permeable shelter belt and over 20 per cent gain in sheep liveweight. Benefits like these have yet to show up on the demonstration farms, since the trees are not yet large enough.

Plant Water Use

The most serious land degradation problems in the demonstration farm region are dry-land salinity, water erosion and tree decline. Improved farm layout, pasture establishment, water supply and revegetation have played a major role in dealing with each of these problems on the demonstration farms.

It is much too soon, however, to comment on the impact of the Potter Plan works on water tables, and consequently on dry-land salinity. Water-table monitoring programmes have been implemented (see Chapter 6), but it will be many years before the effects of the salinity control measures become evident.

One of the selection criteria for the demonstration farms was that they should be high in subcatchments, in order to maximise the opportunity to lower water tables through increasing plant water use. At 'Helm View', 'Cherrymount', 'Daryn Rise' and 'Reedy Creek', the project has the best opportunity to limit the spread of saline areas in the long term, as they are all at the top of their respective watersheds. If groundwater systems are localised, these farms can control their own destiny, rather than being influenced by the land-use practices of farmers higher in the catchment.

Water Supply

The short-term benefits of improved water quality on grazing properties have already been mentioned and the Potter approach to farm water supplies has many long-term benefits. The main features of this approach are:

- natural watercourses such as creeks and rivers are fenced out and re-vegetated with indigenous species;
- watering points such as troughs are located in well-drained areas least prone to erosion;
- large dams or tanks are sited high in catchments (fed by catch drains where necessary) rather than in gullies;
- dams are fenced from stock and surrounded by indigenous shrubs and trees.

These measures have production benefits because they ensure water of reliable quality and quantity and they also confer environmental benefits which help both the farmer and the community. The conservation of wetlands and natural watercourses is critical to ensure the quality of our scarce water resources, to maintain the extraordinary biological diversity of riparian environments and to provide habitat for migratory birds and insects.

Wildlife Habitat

Native wildlife colonised the revegetated areas on the demonstration farms rapidly, as mentioned earlier. The forty species of trees and shrubs planted or seeded in the project were nearly all indigenous, with the exception of some species in situations for which the local species or provenances were not suitable. We tried to use local provenances (genetic strains) wherever we could—for example Red Gum, *Eucalyptus camaldulensis*, is the most widely distributed eucalypt in Australia, and the Murray River or Broken Hill provenances are very different from the plains provenances of south-western Victoria. On the Potter farms, wildlife habitat was provided in a matrix of patch and corridor. Shelter belts and gully revegetation provided the corridors; and mid-paddock clumps, corner plantings, wood lots and dam plantings acted as patches. Few plantings occurred purely to provide wildlife habitat; however most plantings did so because the species planted reflected natural associations.

There is little research data to prove that provision of habitat for native wildlife improves farm production and although demonstration farmers are noticing more birds on their farms, it is too early to expect to measure any effect from improved habitat values. To that extent it is an act of faith to believe that preserving viable habitat for diverse communities of native

- Two approaches to combining farm timber production and shelter. The photograph on the left is an agroforestry site at 'Satimer', where timber trees have been planted in triple rows, with strips of pasture in between. The photograph on the right is of a 'Reedy Creek' wood lot of mainly durable native species, which protects a formerly eroded hill and also provides emergency shelter.

- These two photographs illustrate the fact that shelter belts do not have to be linear. With some planning and attention to stock movement and land type, effective shelter can be established in whatever configuration best suits.

• This classic old western Victorian shelter belt of *Eucalyptus cladocalyx* represents a tremendous farm timber resource. However, it now requires coppicing from one side to provide more low shelter. With current firewood and fence dropper prices, this operation should be very profitable.

• Seed collection and storage is important for large-scale revegetation projects. We had the assistance of the Cavendish staff of the Department of Conservation and Environment for seed collection and extraction. This hothouse was ideal to get seed capsules to open quickly. Here Robert East is extracting Red Gum seed.

animals, insects and micro-organisms is beneficial for farm profitability.

However we are finally seeing some maturity emerging in the farm tree movement in Australia. Farmers no longer feel self-conscious about protecting or re-establishing native vegetation for its intrinsic natural values, nor that every improvement on a farm has to be justified in purely economic terms. It is a basic principle of ecology that maintaining biological diversity is fundamental to the stability of any ecosystem. Despite a lack of evidence to support the production benefits of establishing wildlife habitat (although there is plenty of evidence of the adverse consequences of reduced diversity), an increasing number of Australian farmers is determined to preserve remnant habitat areas and to create new ones, and the Potter Farmland Plan demonstrates how this can be done productively.

Diversification of Income

The main opportunity for diversification of income created by the programme is through the selective harvesting of trees and shrubs. The potential for income from timber, fodder, honey, flowers and essential oils from a farm revegetation programme has already been described in Chapters 3 and 4. These products can be considered in two ways—they can be regarded as a by-product of revegetation, a bonus if they ever yield a cash flow, or they can be regarded as a commercial proposition, an addition to the product range of the farm.

The project has taken the former position in the main, establishing trees and shrubs for reasons other than harvesting for particular products. However, in many of the shelter belts we included one row of species with known commercial value such as Red Gum (*Eucalyptus camaldulensis*), Spotted Gum (*Eucalyptus maculata*) or Red Ironbark (*Eucalyptus sideroxylon*), planted at close spacings so that they can be thinned at the age of twenty or thirty years for posts, poles, firewood or whatever products the market demands and the trees can provide. Similarly, several gully and hilltop plantings, in particular on 'Reedy Creek', 'Cherrymount', 'Daryn Rise', 'Ballantrae', 'Satimer', 'Helm View' and 'Warooka', are wood lots which include Blackwood (*Acacia melanoxylon*), Sydney Blue Gum (*Eucalyptus saligna*), Swamp Yate (*Eucalyptus occidentalis*), and several other eucalypts, casuarinas and wattles suitable for firewood. We also established a demonstration wood lot of *Pinus radiata* at 'Reedy Creek' and an agroforestry planting of eucalypts, pines and Douglas Fir at 'Satimer'. It is likely that all of these plantings will generate cash for the landholders eventually, but their main *raison d'être* is that they assist the operation of the demonstration farms in other ways—any cash return from sale of produce is a bonus.

IMPROVED CAPITAL VALUE

The influence of the Potter Farmland Plan on the cash flow of the demonstration farms is discussed in more detail in Chapter 6. A computer model which generated a fifteen-year budget for a hypothetical farm, based on data from four of the demonstration farms, showed that the key variables which determine the profitability of an investment in whole farm planning are the cost of the works, the degree to which stocking rates are increased and the improvement in capital value resulting from the works.

The consensus of the great majority of those who have visited the demonstration farms, including a professional valuer, is that the value of the farms increased immediately, by at least as much as the implementation cost of the plan, including labour costs. We believe that the value of the Potter farms will continue to increase in relation to that of the farms around them as awareness of land degradation increases and buyers become more discerning. In the last few years there has been a growing tendency in western Victoria for well-treed properties to bring higher prices than expected as landscape values play an increasingly important role in determining price.

● BENEFITS OF A WHOLE FARM APPROACH

So far we have discussed the short- and long-term benefits of discrete elements of the whole farm planning process on the Potter farms. But to examine each element on its own is to lose sight of the integrated nature of the approach which is the distinctive feature of the planning process. The approach we evolved integrates short-term and long-term production goals and the needs of the farmer with the needs of the land, and it ensures that individual farm plans are compatible with regional catchment guidelines. The whole farm planning approach encourages the farmer to recognise off-farm costs and benefits and to plan for the next generation of land users. These features of the process are less tangible and more difficult to quantify but just as real to the farmers who have experienced them.

INCREASED AWARENESS OF ECOLOGY AND KNOWLEDGE OF THE FARM

The first benefit of whole farm planning which most of the demonstration farmers mention is the extra knowledge they have gained about their land. It is clear that the extent to which they have improved their knowledge of their land is directly proportional to the extent of their involvement in the process of redesigning their farms and implementing the farm plan.

The whole farm plans for farmers on whose properties there were only limited works programmes were completed later in the project because of time limits. These plans did not go through as many drafts as those for 'Satimer', 'Gheringap', 'Helm View', 'Willandra', 'Wyola', 'Daryn Rise' and 'Reedy Creek'. Consequently the farmers were less involved in planning and did not have the same opportunities to 'evolve' with their farm plans. In retrospect this was a mistake. It would have been better if all the participating farmers had been through the planning process together, before any works were implemented on their farms in the kind of short courses developed later in the project and described in Chapter 7. However we did not have enough time to do this and, in many ways, the fact that the planning process was modified as it was applied to real farms enhanced its effectiveness and acceptance by the farmers. Logically, of course, the plan should be complete before works are started, but it is often thrown into relief and consequently easier to complete after 'hands-on' experience in implementing the obvious first steps. The fact remains, though, that the farmers who have put the most time into their plans have learnt the most.

It was also obvious by the end of 1986 that the more farmers contributed to the works on their farms the greater their subsequent 'ownership' of the works and the plans upon which they were based. The farmers who had the highest proportional inputs were Peter Waldron, the Milnes and Ross Kitchin. These farms are now the most effective as demonstrations of whole farm planning. That is not to say that the other demonstration farms are not effective, but when the farmers were committed to a major programme of rethinking farm layout and management and 'taking over' the planning process for their farms, their knowledge of the farm, willingness to change, and confidence and ability to implement plans was soon greatly increased.

The lessons from this experience are obvious. Land-use plans, whether based on regions, catchments or individual farms, need to be 'owned' by the people who have to implement them. When such ownership is developed, land users need few inducements to act.

We found that the attempt to define new management units based on land types forced the farmers to re-examine each hectare of land on its merits, rather than looking at the land in 'blocks' defined by existing fences. For the farmers to draw a line on an aerial photo along the boundary between two soil types required intimate knowledge of the farm, and gaps in knowledge were quickly exposed. When these gaps were filled in by inspection on the ground, other more subtle differences in land type (for example differences in pasture composition) were often noticed. By this stage in the planning process, the farmer was already looking at his land in a different way. Peter Waldron commented, 'I used to look at a fence to see whether it needed straining. Now I look at it wondering if it is in the best place.'

As our planning programme continued, some farmers' knowledge about land degradation such as erosion and tree decline increased. For example, it was quite a shock to those in the Red Gum country to contemplate a landscape without majestic old trees within two generations, as is likely at the present rate of decline. All the demonstration farmers learnt more about how to control land degradation. While preparing a whole farm plan, they realised that controlling land degradation and improving short-term farm production are not mutually exclusive. The awareness that ecology is an inextricable part of farm management caused the farmers to look at the land with fresh eyes, with a deeper understanding of the relationship between its management and its long-term ecological stability and resilience and, consequently, their own productivity and profitability.

Many of the Potter Farmland Plan farmers' comments made it obvious that they are now much more aware of the landscape around them and of the need for all farmers to act. According to Peter Waldron, more needs to be done:

> While still holding similar views now as before the project, I think my concerns have been reinforced. I am worried about the future; much grazing and cropping land will be lost before we reverse decline. We have reversed degradation on 'Willandra' but not enough is being done around us. Salinity may still affect us, more trees are needed.

In Leigh Heard's words.

> It has been tremendously exciting to be involved in the Potter Farmland Plan on 'Ballantrae' and with the project in general. From the knowledge and experience we have gained it will be easier for us to continue this work until our vision for the property is fulfilled. I suppose you could also say that the project has also given us more resolve to get on with the job and be looking ahead. There is obviously a tremendous awareness in the rural community of the value of tree planting due to a happy coincidence of factors of which the Potter Farmland Plan is only one. The 'whole farm planning' approach is, however, unique to the Potter Farmland Plan and this is the strong point of the programme which should not be glossed over but is *central*. Sometimes I feel despair driving around the countryside, at how much work there is to be done in tackling the land degradation/farm productivity problem, then over the next rise is a well-planned property that is obviously producing well and that restores my faith! The future? We all must be optimistic!

According to Jeremy and Jill Lewis:

> The project has made us more aware of the countryside—what is happening here and what is not happening in many places.

They now have the knowledge and skills to make significant improvements on their own land and they realise that others could do the same if their priorities changed, so they are in a good position to influence others with confidence and credibility.

One important finding of the project is that the participants are now able to put a whole farm plan into practice much faster than they would have thought possible before their involvement in the project. We estimated at the start that the work on the key demonstration farms was carried out at least three times faster than a landholder would have done it without help; That is, the Potter Plan farmers did in three years work which would have taken ten years on most properties. Furthermore, the landholders who have had the most involvement with the project are now saying that with the experience they have gained, they can now justify a much faster rate of implementation in terms of time and money, even after Potter Foundation funds are no longer available to them. The amount of work done on these farms since 1987 testifies to this commitment and confidence.

In Gavin Lewis' words:

> Prior to the Potter Farmland Plan, management was not as easy, but now with increased fencing and fencing to soil types the dry sheep equivalent per hectare has risen quite a lot. My view has changed a lot in as much as I now feel as though I know how to at least start to arrest some of the problems that prior to the plan were there, but I didn't know how to go about starting a project like this.

A comprehensive whole farm plan was prepared for 'Cherrymount', which Dave McCulloch describes as:

> ... a marvellous idea and gives us an aim. The Potter Farmland project has been a tremendous injection of enthusiasm and improved techniques which has given us direction for this type of improvement.

According to Stuart Cuming:

> My view of farming has been enhanced by the Potter experience. I think it has provided an excellent model for other farmers to take inspiration from... The increased subdivision of large paddocks into smaller ones has helped our management of the stud sheep considerably as there are now nine paddocks where there were four paddocks... I am not convinced that we have really solved our salinity problems by fencing and planting salted flats, more research

needs to be done. Also we are definitely creating a lot more work for ourselves by planting species which blow over and have a short life ... We have only just started to scratch the surface in our knowledge of what has to be done. The Australian landscape is a disaster with most problems getting worse before they get better ... The Potter Farmland Plan has had a profound effect on 'Fernleigh' and on my life and is undoubtedly one of the best things to happen to the Glenthompson area.

The Potter Farmland Plan did not invent whole farm planning, nor did we develop new ways of tackling land degradation. The information used by the project was already freely available and good farmers had been applying aspects of the Potter approach long before the project began. For some of the Potter farmers the changes brought about by the project, to their farms and their thinking, were relatively minor. For example, John Lyons says that the project:

> ... has not changed my views greatly. Involvement with the project has shown us that trees are easier to establish than previously thought. The tree planter and community groups overcome the old labour problem. The emphasis on trees by the project is out of balance with the production required by farms to remain viable. Capital appreciation is only a nice feeling if one sells out. Improved pasture species and management techniques will lift production, trees will improve the shelter aspect but in many cases drainage will be the only way excess water is removed from the surface to prevent waterlogging. Increased pasture production slows the natural drainage, maybe even increases the problem of waterlogged soils.

There is no such thing as a *best* farm plan. The Potter Plan farmers found that one of the main advantages of the planning process was its flexibility. As described earlier, the plans were changed (and are still being changed) as we came to understand each farm more fully and as the landholders became more involved in the process and saw the success of some of the early works. You would need different plans for a given farm for different farmers and for different enterprises; the common factor in the different plans would be harmony between farm development and the ecology of the land.

As the whole farm planning process evolved, it became obvious that the materials required for farm planning are simple and cheap. The most important ingredients required to prepare a good plan are an intimate knowledge of, and a willingness to learn more about, the land; an open mind; and a desire to live *with* the land in a stable, balanced relationship, rather than *from* the land in a manner which cannot be sustained in the long term.

Sustainable Productivity

Whole farm planning is essential if Australian farmers are to change to more sustainable land use. Integrated planning means that farmers can design and manage their farms to ensure that short-term production goals are not achieved at the expense of the future productive capacity of the land. The approach identifies the different types of land on the farm, determines the limitations and capabilities of each land unit and plans the management of each unit accordingly. Parts of the farm where external inputs are not justified are identified and managed accordingly. Properly applied, whole farm planning encourages farmers to work with the land rather than against it, ensuring that the land will be used in a more sustainable manner than has been customary. Farmers can manage each land unit in appropriate ways, which makes for maximum efficiency. In this manner, the approach draws out the needs of the people on the land and forces them to relate these needs to the capacity of the land to meet them.

More Effective Use of Outside Advice and Extension Services

Another benefit of whole farm planning is that the plan provides a focus for distilling advice from a range of different sources. The plan is a framework for gathering, synthesising and applying information. Farmers are the ultimate generalists, incorporating skills from animal husbandry to engineering, personnel management and accountancy, yet they tend to receive advice from a very diverse range of specialists, as well as from other farmers. Fleshing out the skeleton of a whole farm plan can help to show where the various jigsaw pieces of specialist advice from a range of disciplines can be used in relation to each other and to the overall goals of the farm and the people relying on it. Compromises often have to be found, for example between what the accountant or bank manager would prefer according to the demands of next year's tax return and what the wildlife extension officer would prefer according to the habitat requirements of a particular native animal.

Armed with a developing farm plan, a landholder can use advice from extension officers more effectively and ask more appropriate questions to elicit a more useful response. For example, the precise question 'How can I increase my plant water use and still maintain or improve stock numbers?', is much more likely to elicit useful information than the general, 'What can you tell me about salinity control?'

During the weeks spent preparing plans with CFL and DARA staff, we found that it was a valuable exercise for the farmers to have two advisory officers from different disciplines work together on a plan. Non-professional advisers, such as other farmers, were often more comfortable when given a specific question related to a plan in front of them, rather than being asked a general, abstract question which was difficult to apply to the concrete.

As we worked it became obvious that our integrated approach to whole farm planning, which drew together the needs of various parts of the farm enterprise, the aspirations of the farmer and the ecological needs of the land, ensures the best use of existing information and advice. It is also likely to reveal the areas where information and advice are lacking.

CONTINUITY OF MANAGEMENT

Most of the Potter Farmland Plan farmers want to hand their land on to their children as a viable asset. On most farms this 'handing on' happens gradually as decision making shifts from one generation to the next. On 'Daryn Rise', 'Ballantrae' and 'Wyola', the younger generation are already actively involved and on 'Helm View', 'Warooka', 'Reedy Creek' and 'Nareeb Nareeb', the older children are at tertiary institutions, either preparing for careers on the land or gaining skills in other areas to give themselves alternatives. In all cases the whole farm plan provides an opportunity for parents and children to make long-term decisions together.

Where more than one person is managing a farm, a whole farm plan is a useful basis from which to explore the goals and potential inputs of each person. This is especially useful when two or more generations of a family run a farm—whilst it does cause arguments, preparing the plan exposes the views of all family members. One Hamilton district farmer bought a large-scale aerial photograph of his farm for each of his sons and daughters and asked each of them to prepare a whole farm plan. He used the results not only to evaluate their interest, thinking and future goals for the farm and to give him new ideas, but also to help determine whom he should leave the farm to!

Simply having a plan on paper helps. As Peter Waldron often says to groups visiting 'Willandra', 'I had a farm plan in my head for years—which was a bloody stupid place for it!' With an aerial photograph and coloured overlays, the whole family can see what the long-term goals for the farm are, and often feel more able to offer ideas and suggestions.

When families or other partnerships prepare a farm plan together, the plan develops as people's knowledge, confidence and circumstances change, just as plans prepared by individuals do. When management transfers from one partner to another, however, the existence of a plan assists continuity in that the contribution of the previous manager is recognised and available for consultation.

FRAMEWORK FOR FARM AND PERSONAL GOALS

Just as a whole farm plan provides an excellent focus for advice from various sources, it is also an excellent framework in which to reconcile the development of the farm with personal and family goals.

When farmers begin to prepare a whole farm plan, and as the plan develops, their natural tendency is to try to imagine what the farm would look like when the plan is completely implemented in twenty or thirty years' time. This inevitably leads to the consideration of personal and family goals, which must then be built into the plan—an exercise which seems to happen only rarely on family farms. An example from one of the demonstration farms illustrates this point. Greg Milne, son of Lyn and Bruce, has worked hard implementing the Potter project works on 'Helm View'. In 1986 he said that if he had one wish, it would be to go forward thirty years, just for one day, to see what the farm would look like, and that would be enough spur to keep him going in the meantime. In saying that, his own long-term goals emerged and the incentive to implement the plan was reinforced, not just for him but for the rest of the family too.

Some questions which should be asked in the early stages of farm planning provide a useful framework for reconciling personal goals with farm development. In the first stage of preparing a whole farm plan, it is appropriate for farmers to reflect for a moment on the way in which they make a living from the land and on the implications this has for future management of the land. Most people ask themselves at some time in their life, 'Why am I earning a living this way?' Trying to answer this question honestly on a farm is a very worthwhile exercise.

Farmers' long-term options include:

1   To retire and sell.
2   To continue to improve the farm and hand it on to the family.
3   To continue to improve the farm and sell.
4   To diversify and stay on the farm.
5   To get out and try something different.

In considering the present enterprise on the farm in more detail, we found the following questions to be useful 'thought starters' in whole farm planning short courses;

1   Do you consider that your present enterprise is one that the land can sustain indefinitely?
2   Does it provide you with a satisfactory income and lifestyle—both now and in the future?

3 Could you diversify into another enterprise?
4 Could you add value to your products or seek a more favourable market?
5 Have you tried to find out what the local and overseas markets want?

It is not good enough for the farm-planning process to simply start with an analysis of different land types or some thoughts about where fences should be located. When the questions above have been considered by the farmer and all others potentially involved in the management of the farm, they are in a much better position to consider the management of the land and to commit themselves to implementing a programme of long-term improvements.

These questions were used in whole farm planning short courses we ran in the Hamilton area in 1987 and 1988 (see Chapter 7) to great effect. Perhaps the people who are prepared to pay $150 and give up six half days for a short course are a biased sample, since they must already have a long-term interest in farming, but in general the consideration of personal goals and farm improvement goals strengthened the courses and the commitment of the participants to their plans.

[1] Steve Howett and Andrew Lothian, *Benefits of Tree Establishment on Farms. A survey of landholders' experiences*, Department of Environment and Planning, South Australia, 1988.
[2] J. W. Cary, A. J. Beel and H. S. Hawkins, *Farmers' attitudes towards land management for conservation*, School of Agriculture and Forestry, University of Melbourne, 1986.

6

Keeping Track

● MAINTAINING RECORDS

O PERATIONS RECORDS were compiled and stored on copies of sections of the farm plans, prepared by John Marriott. One of these diagrams is illustrated on page 98. The diagram shows a typical paddock plan traced from the enlarged aerial photographs on which the whole farm plans were based. For each operation on each paddock on each farm in each year of the project, data was entered on similar plans under the following headings:

Fencing

- Length of each section;
- number of end assemblies;
- fence design;
- quantities of wire, posts, ties, insulators, droppers required;
- personnel and number of person hours involved if a contractor was used.

Site preparation

- Date of works;
- type of operation, such as ripping or spraying;

- herbicides, equipment and application rates;
- personnel and number of person hours involved;
- weather and soil conditions.

Planting/seeding

- Date;
- personnel;
- type of operation;
- number of plants or quantity of seed of each species;
- equipment used, number of hours involved;
- rates per person hour of planting and guarding;
- type of guards;
- weather and soil conditions.

Maintenance

- Date of each maintenance operation;
- type of operation—for example, replants, weed control, slashing, guard removal;
- number of plants of each species required;
- herbicides, equipment and application rates;
- personnel and number of person hours involved;
- weather and soil conditions.

● MONITORING IMPACTS

The Potter Farmland Plan is not a research project. It was set up to demonstrate how farm design and land management can be reorganised so as to achieve more sustainable production. Demonstrations are more effective if they can be accompanied by data showing the influence of the measure being demonstrated. Whilst one of the features of the whole farm planning approach employed on the demonstration farms is its holistic nature, we monitored the influence of particular aspects of the approach as well as that of the project as a whole.

VISUAL RECORDS

An extensive photographic record of the programme was compiled. It comprises 5500 35-millimetre colour slides taken by project staff and several hundred colour transparencies taken by DCE photographers Noel Ryan and Rawden Sthradher. These photographs are catalogued by farm and by subject material—for example 'farm layout', 'land degradation', 'public relations'.

GROUNDWATER LEVELS AND SALINITY

Groundwater levels and salinity are being monitored on 'Daryn Rise' at Glenthompson and 'Willandra' at Melville Forest. At 'Daryn Rise', DCE and DARA are conducting a groundwater and plant-water use research programme funded by the State Salinity Bureau, on three very similar adjacent subcatchments, two of which are on 'Daryn Rise'. Potter Farmland Plan works have taken place on the western half of 'Daryn Rise', and the eastern half is completely devoid of trees. Peter Dixon of DCE is supervising a programme which aims to establish a vigorous perennial pasture over the entire treeless subcatchment. There is an intensive network of piezometers over the sub-catchment, in order to monitor and quantify the influence which pastures alone can have in increasing plant water use, hopefully reducing accessions to groundwater and consequently the level of the water table. The pastures-only subcatchment will be compared with the trees and pastures approach on the Potter Plan subcatchment, and the unimproved subcatchment (also treeless, but with native annual grasses only). Notched weirs have been established in two of the subcatchments to enable run-off to be measured.

The three catchments have also been surveyed by Dr Baden Williams of CSIRO in Canberra, with an electromagnetic detection device which can map the distribution of salt in the soil profile. This is a much faster technique than drilling to establish piezometer nests and it does not involve digging holes. In this case the two techniques can be compared to further calibrate the electromagnetic mapping.

Peter Waldron initiated his own network of piezometers on 'Willandra' to see what was happening to his water table and to assist in quantifying the influence of Potter Farmland Plan works. Like the data from the 'Daryn Rise' bores, Peter's water table readings are sent to Bendigo for storage on computer, with DCE's other piezometer data from around the state. Baden Williams also mapped the distribution of salt in the soil profile at 'Willandra'. The piezometers at 'Daryn Rise' and 'Willandra' will enable monitoring of the influence of revegetation (including improved perennial pasture) on groundwater levels and thus determine the degree of success in tackling dry-land salinity.

REVEGETATION

It is important that the Potter Farmland Plan demonstrates the 'state of the art' in farm revegetation, using the most efficient and effective techniques in tree establishment and protection. Final-year Forest Science students from the University of Melbourne assessed the success of the 1985 and

1986 plantings and seedings in December 1985, January 1986, May 1986, December 1986 and January 1987, and carried out a partial assessment in May 1987. These confirmed that the establishment techniques used for tube stock plantings were successful, as the table on page 94 illustrates.

It was more difficult to evaluate the success of direct seeding. Small germinants were hard to find and some species seemed to germinate as late as eighteen months after seeding. In many cases when initial surveys suggested that seeding had failed, germination was found to be quite satisfactory one or two years later. The direct-seeding operations were not designed to be statistically valid trials, and results were so variable that it was impossible to draw conclusions which would have altered our 'best-bet' establishment regimes for direct seeding.

More systematic direct-seeding trial work has since been undertaken in a research programme funded by the National Soil Conservation Program and directed by Dr Rod Bird and Keith Cumming of the Pastoral Research Institute at Hamilton. This study involves four regions: the southern Wimmera; at Glenthompson on Ordovician sediments; on basalt soils south of Hamilton and on 'Willandra'. Peter Waldron is convinced that direct seeding is the best establishment technique and has developed his own method, as discussed in Chapter 4. The best examples of direct seeding in the project are now at 'Willandra'. These farmer-directed examples will make an excellent comparison with the controlled experimental work, which has been replicated each year for three years to allow for seasonal fluctuations, with outstanding results overall.

The Potter Plan tube-stock plantings are complemented in the Melville Forest and Wando Vale areas by agroforestry trials carried out from 1985 to 1987 by Rod Bird and Keith Cumming. Wood lots were planted on six farms to investigate the performance of forty different species with commercial value, including eucalyptus, casuarina, allocasuarina, acacia and pinus species. The Melville Forest and Wando Vale sites include spacing trials which will investigate the influence of tree density on tree and pasture growth. The Wando Vale site is on 'Barnaby'. As for all demonstration farm plantings, accurate data on establishment details has been recorded. The combined work of the Potter Farmland Plan and the Pastoral Research Institute, both in direct seeding and tube-stock planting, provides an excellent source of information about which trees to grow in western Victoria and how to establish them successfully.

CASH FLOWS

It became obvious during the second year of operations that we needed to assess the economic impact of the project works on the demonstration farms. We looked at four farms in detail—'Daryn Rise', 'Gheringap', 'Helm View' and 'Willandra'—to examine all cash inputs associated with the Potter Farmland Plan works, and initial cash outputs. Using the data from these four 'real' farms, and from the forty farms of the DARA south-west monitor farm project, which has been running since 1970, we devised a hypothetical 550-hectare farm as a model, with the assistance of Andrew Patterson, the DARA Farm Management Economist based at Hamilton.

We worked out two different budgets for the model property for fifteen years—one assuming that a whole farm plan was implemented during the first three years and one assuming no expenditure on a whole farm plan. These cash flows were then converted to Net Present Values using real discount rates of 5 per cent and 10 per cent, which correspond respectively to 12 per cent and 17 per cent nominal, assuming an inflation rate of 7 per cent. The Net Present Values were then compared to ascertain the profitability of the whole farm plan over fifteen years.

The major assumptions differentiating between the whole farm plan farm and the 'do nothing' farm are summarised below:

Differential Assumptions		
Assumption	Whole farm plan approach	'Do nothing' approach
Non-arable area	30 ha over 3 yrs, then zero	5 ha, then 1 ha/year increase
Added capital value	$100 per ha after 3 years	zero
Development costs	$50 000	zero
Labour saved	2 weeks per year	zero
Carrying capacity	10.5 d.s.e.*/ha after 3 yrs	10 d.s.e./ha
Shelter	zero	zero
Timber income	zero	zero
* dry sheep equivalents		

Using the assumptions above, the difference in Net Present Values (profitability) between the 'do nothing' farm and the farm implementing a whole farm plan was calculated. According to the model, the whole farm plan 'pays for itself' after six years if the discount rate is 5 per cent, or after eight years if the discount rate is 10 per cent. It is unlikely that any real situation will be identical to the one in the model, so it is important to determine which assumptions contribute most to the profitability or otherwise of investment in a whole farm plan. We varied each assumption in the computer model while all other assumptions were held constant and calculated profitability at each point to test the sensitivity of the model to each assumption.

The sensitivity analysis revealed that the factors which have most influence on the profitability, over fifteen years, of implementing a whole farm plan on a 550-hectare sheep and cattle property in western Victoria are: the increase in capital value, the cost of implementing the whole farm plan and the effect on carrying capacity.

Each of these assumptions was very conservative. We assumed that, after redesigning farm layout, re-establishing pastures on saline and eroded areas and creating extensive shelter belts and wood lots, the stocking rate would only increase enough to maintain total stock numbers. Several of the demonstration farms have in fact increased stocking rates after three years but, with seasonal fluctuations and a lack of hard data, this is a subjective observation which requires further analysis.

We also assumed that the increase in capital value would only be equal to the money spent, yet the subjective opinion of visitors to the demonstration farms is that their market value has increased significantly more than the total expenditure on works. This is backed up by a study carried out by Jack Sinden of the University of New England, who found that expenditure on soil conservation works in the Manilla Shire was reflected by an increase in market price 1.28 times the cost of the works.

The cost of implementing the whole farm plan was taken directly from the demonstration farms and included the cost of the farmer's labour at contract rates for fencing, pasture establishment and revegetation works. Thus, the implementation costs used in the analysis are higher than the actual cash costs which would be incurred by farmers implementing whole-farm plans themselves.

Most farmers do not cost their labour and would only consider cash costs when assessing the profitability of an investment in whole farm planning. Another critical assumption—the increase in capital value—may change from a farmer's perspective. Many farmers do not regard increased capital value as an incentive, saying that they do not intend to sell and that the only effect capital gain has on their cash flow is to increase their rates.

To test the profitability of this kind of 'farmer's costing' of investment in whole farm planning, we changed the model and assumed that there was no increase in capital value and no labour costs. This change showed that, for farmers who choose to ignore capital gain and the cost of their own labour, an investment in whole farm planning is more profitable over fifteen years than if increased capital value and labour costs are included in the analysis. Although the whole farm plan investment takes slightly longer to break even at discount rates of 5 per cent and 10 per cent, with the changed assumptions (eight and nine years respectively, compared with six and eight years), the end result after fifteen years is more profitable.

• Options for simple fencing of clumps. The Whiteheads Creek Landcare Group lean star posts in to prevent cattle rubbing on the fence, which seems to work. We found that the solar-powered small energiser created a very effective short electric fence on this clump at 'Helm View'.

• This sequence illustrates a very versatile lift-up gate on my family farm, 'Crowlands', north of Hamilton. The hinged top rail allows the fence to lie down so that tall vehicles can go over, rather than under the fence. It is still much cheaper than a conventional swinging gate.

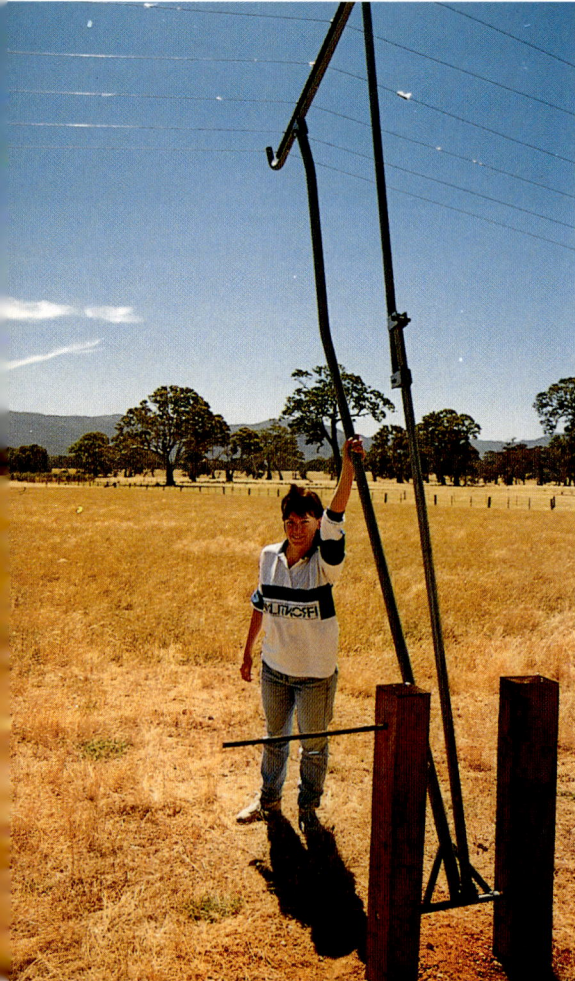

• Understorey species are just as important as tall trees in farm revegetation. We collected seed from the *Bursaria spinosa* above to use on the demonstration farms. Natural regeneration from Red Gums can be spectacular in the right conditions. Simon Lewis of Cavendish established thousands of seedlings (below) by controlling weeds around the mature trees while their seeds were falling.

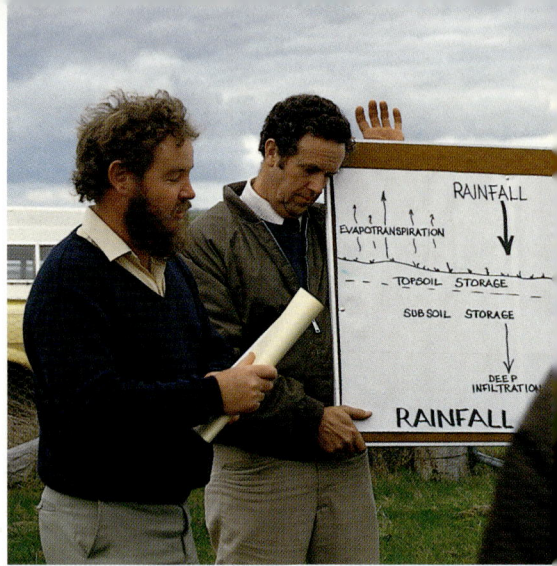

● 'Hands-on' involvement and practical demonstrations are the keys to helping people to understand most things, including the complexities of balancing production and conservation on farms.

It was interesting to note that, in economic terms, the rate of land degradation had little impact on the profitability of the whole farm plan investment and neither did any reduction in lambing losses. Nevertheless, as discussed in Chapter 5, both these factors are cited by farmers in the region as reasons to establish farm trees. This indicates either that land-holders are uninformed, or that they use criteria other than economic efficiency when making decisions about land management.

Our experience suggests that the latter suggestion is correct and that the influence of a whole farm plan on cash flows should be viewed in this light. The analysis described above did not take into account many other factors which would influence landholders in decisions about land use. The satisfaction of living and working in a more attractive environment, the off-farm benefits which affect the wider community, the incentive to leave the land in a better condition for future generations, the desirability of creating habitat for native wildlife, the possible diversification of farm income after fifteen years, and the benefits which accrue after fifteen years were all ignored in the analysis.

Even with the conservative assumptions adopted, however, an investment implementing a whole farm plan in western Victoria is profitable after ten years.

The monitoring activities outlined above will yield valuable information and provide a base for future analyses; however, we did not seek commitment from other agencies to on-going monitoring programmes during the project, because monitoring tended to have a lower priority than the increasingly intensive works and public relations and extension programme. As a consequence, apart from the instances mentioned above, other organisations have not been involved in monitoring the influence of Potter Farmland Plan works on the demonstration farms or on other landholders. To some extent this can be rectified, as the existence of comprehensive baseline data will enable comparative studies between the demonstration farms and others in the future. However the lack of involvement by government departments, universities, agricultural colleges, local community groups and schools in monitoring the Potter Farmland Plan is a major flaw at this stage of the project.

7

Spreading the Word

THE POTTER FARMLAND PLAN is ultimately about getting across a message. As the project developed we spent more and more time refining and disseminating this message. The objectives of our public relations and extension programme from 1985 to 1987 were:

1   to promote the concept of whole farm planning at local, state and national levels;
2   to create an awareness of the existence and features of the Potter Farmland Plan demonstration farms at local, state and national levels;
3   to provide opportunities for guided access to the demonstration farms by farmers, extension officers, farmer educators, the media, policy makers, community interest groups and the general community to see the practical application of the whole farm planning concept;
4   to stimulate the development of similar demonstration activities in other areas; and
5   to publicise and widely disseminate the planning and outcomes of the project.

FIELD DAYS AND FARM TOURS

We organised field days in each of the demonstration areas in late March or early April in 1986, 1987 and 1988. They were designed to give interested people the opportunity to inspect the demonstration farms in each area and to talk to the participating landholders. Apart from field days, on which the public were generally invited to inspect demonstration farms, we organised many tours, usually at the request of an interested group of people.

The tours were intended to give visitors an insight into whole farm planning and an opportunity to meet the farmers on their own land. The farmers' main involvement in the public relations and extension programme was in showing visitors around their own farms. In the early stages, John Marriott or I led the tours but later the farmers began to lead the discussion on their own farms. These tours took visitors through the process of preparing a whole farm plan, using the farmers' experiences to illustrate the benefits of the approach.

The steady increase in the number of tours of demonstration farms is clearly depicted in the graph below. The average number of people per group remained constant, at slightly under twenty.

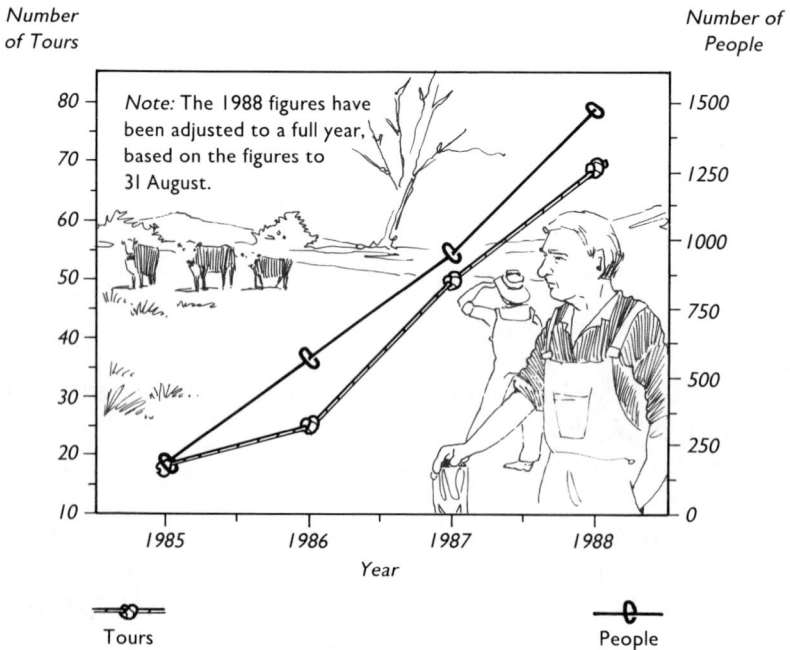

• Inspection tours of demonstration farms, 1985–8

The field days were advertised in the rural press, on local radio stations, in local newspapers and by the distribution of leaflets. In 1987 we also paid for a flyer advertising the field days and addressed 'to the farmer' to be distributed by Australia Post (at a cost of $300 for 3000 letterbox drops) to all farmers and rural businesses in the Hamilton area. To test the effectiveness of this form of advertising, we asked people who attended the field day to fill in a questionnaire describing how they had found out about them. The questionnaire revealed that ABC Radio was easily the most effective form of advertising, followed by the *Hamilton Spectator*, the *Stock and Land*, 3HA Hamilton and other people. The letterbox drop was next, but it was cited by only eight people out of ninety-nine, which seems to indicate that farmers treat 'to the farmer' letters as junk mail. Most of the 254 people who attended the 1987 field days would have received a flyer in their letterboxes.

The field days introduced people to the concept of whole farm planning by examining its application to the demonstration farms. The participants visited each farm, at which the landholder introduced visitors to the farm and the type of operation involved, then the whole farm plan was explained and significant demonstration sites were inspected. Participants did not see every part of every farm, especially after 1987, but all aspects of whole farm planning were covered and the key features of the Potter Farmland Plan approach—integration of production and conservation goals, 'farmer-driven' planning, practicality and innovation in application—were continually emphasised.

Fencing was a major part of our works programme. The three fencing field days and numerous farm tours revealed tremendous interest, particularly from farmers, in the project's fencing techniques.

● SPEAKING ENGAGEMENTS AND MEDIA COVERAGE

As the project developed, the number of invitations for project staff to speak to various groups increased steadily. At the beginning of the project we accepted all these invitations readily because we wanted to maximise awareness of the Potter Farmland Plan, its aims, methods and results. At a typical speaking engagement, we spoke about the background to the development of the project, the need for farmers to rethink their land management to take the whole farm into account and our approach to whole farm planning, before describing the practical activities of the project. A great number of the groups was interested enough in the project to initiate speaking engagements and many of these functions led to inspection tours of the demonstration farms.

The graph below shows an increase in the number of speaking engagements, particularly from 1986 onwards, as word about the project spread and both the project and its staff became better known. The graph shows only functions at which project staff gave prepared speeches; it does not include functions such as 'Hamilton Sheepvention', an annual two-day agricultural 'expo' which usually attracts over 10 000 people, at which we mounted displays and spoke to hundreds of people on each occasion.

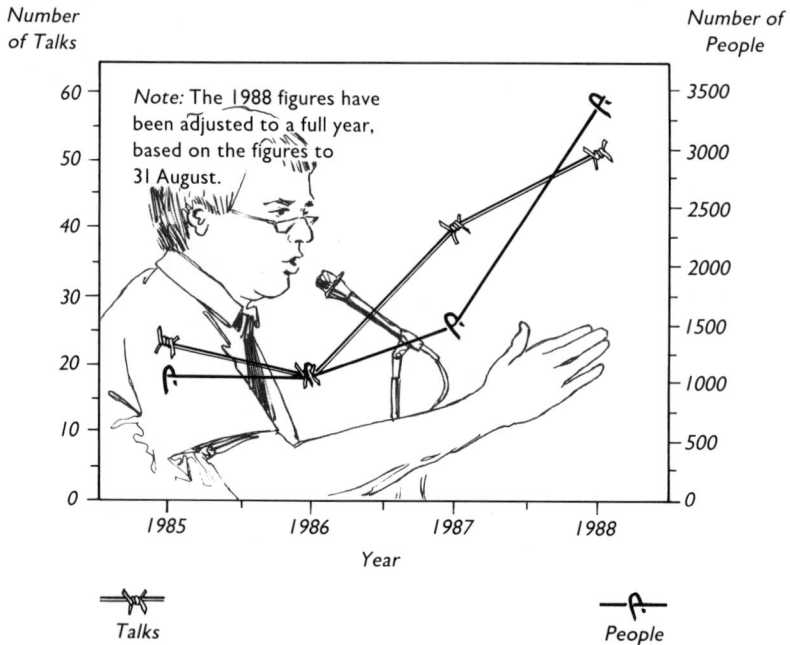

Note: The 1988 figures have been adjusted to a full year, based on the figures to 31 August.

• Public speaking engagements, Potter Farmland Plan staff, 1985–8

In the early stages of the project, most of the groups we spoke to were farmers who had been interested in the objectives of the project, in particular farm trees, for a long time, and scientists from Victoria and elsewhere working in land management, also primarily interested in farm trees. As the project developed, we began to receive invitations to speak to school groups and community organisations such as service clubs, and 'non-tree' farmers at functions such as Victorian Farmers Federation branch meetings and Grassland Society of Victoria seminars.

The increasing diversity of our audiences matched the growth in awareness of the project. The latter was enhanced by media exposure, which is summarised in the table below. We kept local media up to date with progress. The exposure in state, national and international media, for example *National Geographic,* the *Weekend Australian,* the *Financial Review,* 'Countrywide' and 'Earthwatch', was stimulated by contact made by members of the executive or project staff. By 1987, awareness of the Potter Farmland Plan within the rural press was very high, and by far the majority of stories was initiated by journalists rather than project staff. Support from local journalists, particularly ABC radio 3WV at Horsham and the *Hamilton Spectator,* was excellent throughout the project.

Number of articles, interviews					
				Newspapers & magazines	
Year	TV	Radio	Local	State/National	Farmer/Region
1985	2	3	6	5	4
1986	1	4	6	2	5
1987	4	4	9	6	9
1988		3	10	4	10

Note: These figures do not include advertisements, or interviews promoting field days

• ASSOCIATE DEMONSTRATION FARMS

Fifteen demonstration farms were selected from forty-five farms. Many of the farmers who applied unsuccessfully for selection met most of the criteria and supported the ideals of the project. Early in 1985 the Potter Farmland Plan Executive determined to encourage these landholders, termed 'Associate Landholders', to continue their interest in the project.

Although no funds could be diverted to this group, the executive decided to help associate farmers to prepare whole farm plans and to allow them to buy any materials such as fencing materials, seedlings, herbicide, tree guards and pasture seed required to implement their plans at project prices, which were considerably lower than recommended retail prices.

Since the project staff had no time to work out whole farm plans with thirty associated farmers, the executive decided to use private consultants. In early 1986 private consultants offering whole farm planning services were invited to submit samples of their work. ACIL Australia Pty Ltd, Rural Trees

Australia Pty Ltd and Farm Management Consultants Pty Ltd expressed interest in the contract. Three associate farms similar in size and problems were selected and each firm was asked to prepare a whole farm plan for one of the farms. These plans were presented to the local advisory group, members of the Executive and staff from the departments of Agriculture and Conservation, Forests and Lands in June 1986. We decided to ask Rural Trees Australia and ACIL Australia to prepare whole farm plans for the associate farms. They were to offer this assistance to the associate landholders on the same basis as the demonstration farms; that is, the farmer would be expected to meet one-third of the cost of the plan. The local advisory group set down guidelines specifying format and content, so that the plans prepared by the two firms would be comparable.

The associate farmers were asked if they were interested in a whole farm plan prepared with a consultant and advised of the approximate contribution they would be expected to make. Fewer associate farmers than expected expressed interest, citing the cost as a major deterrent, indicating that they would rather attend a farm-planning short course and prepare their plans themselves. Rural Trees Australia (now Rural Planning Australia) prepared plans for the three interested Wando Vale farms and ACIL prepared a plan for a farm at Glenthompson.

● WHOLE FARM PLANNING SHORT COURSES FOR FARMERS

To enhance the impact of the Potter Farmland Plan and to influence agricultural educators, the executive decided in mid-1986 to offer short courses for farmers in whole farm planning. The courses were designed to guide farmers through the process of developing whole farm plans for their own properties, with the help of a full-time coordinator who would use local technical experts in specific areas. The Further Certificate in Farming courses which were run in the Hamilton area by John Sutherland, of the Victorian College of Agriculture and Horticulture (VCAH) Glenormiston, for several years before 1986, provided a good model.

To assist in designing the short courses and to introduce farmer educators to whole farm planning, a four-day workshop was held at Glenormiston in October 1986. The participants were all involved in agricultural education, either in TAFE or other tertiary institutions or in agricultural extension with the departments of Agriculture or Conservation, Forests and Lands.

The first afternoon of the workshop placed the need for whole farm planning in context. Dr Brian Roberts of the Darling Downs Institute of Advanced Education presented a paper describing the extent of land degradation in Australia and the principles of whole farm planning. Dr Max Day of CSIRO presented a paper on the ecology of farmland and the importance of ecological diversity, and Mr Keith Turner of Melbourne University discussed the unique characteristics of Australian soils and the importance of sustainable soil management. In the evening, we outlined in detail the whole farm planning carried out on the demonstration farms, in a session designed as a briefing for the field trip to the farms in the Melville Forest area to take place the next day (6 October 1986). The field trip introduced participants to the demonstration farmers and the practical realities of whole farm planning.

After the first two days, the workshop participants who were interested in whole farm planning endorsed the need for courses and the applicability of the short-course format. The next day and a half were spent attempting to design a course, deciding content, the resources required and the most suitable structure. A steering committee was formed to design a course in detail and Glenormiston agreed to run a pilot course in an area close to the demonstration farms. Participants wishing to run further courses would then be able to learn from the Glenormiston course. The steering committee prepared a manual for course presenters, based on a format of one half day per week for five to seven weeks and another day six months later.

The pilot course was run at Glenthompson from 2 June to 14 July 1987, in conjunction with VCAH Glenormiston. The course was advertised, brochures promoting it were sent to associate landholders and people in the Glenthompson area who had expressed interest in finding out more about the project and an information night was held in the Glenthompson Hotel on 30 April. Approximately twenty-five people attended the information night and nineteen enrolled in the course. This was regarded as a suitable number and some late applicants were turned away.

Resource people from CFL and DARA, from VCAH Glenormiston and from ICI were all used in the course. The course facilitator was Philip Jobling from Glenormiston, assisted by John Marriott and me. People from as far away as Bacchus Marsh, Hawkesdale and Balmoral attended. For a first effort the course was very successful and all the participants indicated that they would recommend a similar course to other farmers.

The most popular topic by far was farm trees, followed by fencing and pastures. Notes were handed out by each resource person, including a *Whole Farm Planning Manual* which outlined the process of preparing a whole farm plan, drawing together the separate aspects of layout, revegetation, water supply, land degradation control and farm management improvements. The experiences of the resource people in conducting the pilot course were reviewed by another meeting of the participants at the Glenormiston workshop and suggestions for further courses were discussed.

In response to demand from farmers, two further short courses were run at Melville Forest and Wando Vale in March and April 1988, once again as a joint operation between Philip Jobling of VCAH Glenormiston and John Marriott of the Potter Farmland Plan. The courses ran concurrently, with a morning session at the Melville Forest Hall and an afternoon session at the Wando Vale Hall. This made most efficient use of the resource people, ensuring that each person was only required for one day. The response to these courses was very favourable.

The demand from farmers in the Hamilton region for farm-planning short courses continued after the Potter Farmland Plan Field Days in March, so further courses were organised and conducted at the Hamilton Institute of Rural Learning in May and June 1988.

After organising and conducting five courses in conjunction with VCAH Glenormiston, some key points emerged:

1   It is important to 'take the course to the farmer', rather than run courses in centralised education institutions. This enables courses to be tailored to the particular requirements of a district, builds the sense of community 'ownership' of the course and minimises travelling time and problems with school buses, for example, for the participants.

2   Interaction between participants in the course is at least as important as interaction between participants and the facilitator and resource people, since it shows farmers that others are confronting the same issues, constraints and problems as themselves. The issue of land degradation in particular focuses attention on the need for a cooperative effort by all landholders within regions or catchments. When neighbours participate in a farm-planning course together, boundary issues have to be discussed and the interaction which ensues is fascinating to observe. Farmers are usually more prepared to make radical suggestions about farm improvements a neighbour should make than about their own farms. Most participants relish the chance which local courses provide for each person to have an input into the plans of other people from the local area. This interaction and the friendly competition which often develops seem to act as catalysts in the conceptual development of farm plans.

3   It is important that at least ninety minutes in each session be devoted to work on the farm plan, so that all participants (assuming that the ratio of participants to resource people is less than 10:1) have time to

question resource people individually about their own plans. For this reason, resource people's presentations should be kept short.

4   The course facilitator must be sufficiently familiar with the district to know the local resource people well enough to coordinate their inputs and cut short their presentation time if necessary. As the facilitator is present at each session, it is valuable if he or she has a good knowledge of the farm-planning process and is familiar with land-use issues in the district in which the course is run.

5   An essential first step is for the participants to introduce themselves to the group in turn, outlining the type and history of their farm operations, the problems they are facing and what they want to get out of the course.

6   If the resource people are all local, it should be possible for them to meet before the course to discuss how the information they provide fits into the course and to ensure that it is in the local context. It should also be possible for participants to follow up issues, either during or after the course, with local resource people.

7   At least one participant should volunteer or be picked at random during each session to describe progress on their own plans and to invite inputs from other participants. Anyone with a farm-planning problem they would like to discuss with the group should have the opportunity to do so.

8   All participants in the Potter Farmland Plan courses to date have asked for follow-up advice on their own farms. This could be built into courses as an option subject to an additional course fee, or, if this option is not feasible, there should be a reunion of participants, six to twelve months after the course, to review progress on the plans, to examine any problems which have emerged and to provide a catalyst for further development of plans.

● PUBLICATIONS

The material written by project staff from 1985 to 1987 had two main objectives. Firstly, to increase awareness of the Potter Farmland Plan and, secondly, to provide detailed explanations of the principles and techniques used on the demonstration farms. The publications whetted the appetite of those who received the small number of copies available from the Hamilton office, but their limited distribution to date has not achieved the original objective to publicise the planning and results of the project widely.

● AUDIO-VISUAL PRODUCTIONS

During the active period of the project, from 1985 to 1987, we made several short audio-visuals for general extension use. However, it became clear to the Potter Farmland Plan Executive that it was not possible to develop short courses quickly enough to cater for farmer interest in the project and whole farm planning. When the not-for-profit company, Potter Farmland Plan Ltd was formed in 1988, one of its key aims was to develop a series of videos on the project and whole farm planning to assist farmers to prepare whole farm plans all over Australia, hopefully applying the Potter principles to their own situations, referring to local specialists for advice where necessary. One of the key target audiences from the start was seen to be the rapidly growing number of landcare groups.

There is a plethora of extension videos available to farmers, so it was important that the whole farm planning videos be of very high quality and have enough information to make them really useful to farmers. During 1989, John Jack and Pat Feilman worked tirelessly, raising sponsorship from BP, Western Mining Corporation, Greening Australia, the R. E. Ross Trust, the Ian Potter Foundation, the *Weekly Times*, ANZ, Union Insurance, Telematics Course Development Fund Trust and Commercial Union Insurance, to ensure a quality production. Richard Keddie was commissioned to write the script and produce the videos, Neil Inall was engaged as presenter, and The Film-house was the production company.

Filming took place in late 1989 and early 1990 on the demonstration farms in western Victoria and on outstanding farms in South Australia, Western Australia and Queensland. The series, called 'On Borrowed Time' was put together in eight parts on two 100–minute VHS cassettes. It outlines the whole farm planning process and its relevance to sustainable agriculture, describing farmer case studies, largely using the farmers to tell the story. It is distributed by the Victorian Conservation Trust and sold for $160 in 1990, but landcare groups have been subsidised by the National Soil Conservation Program.

8

Wider Impacts

B Y 1988 the level of awareness of the Potter Farmland Plan was very high, locally and at state and national levels. In the Hamilton area it often seemed that the profile of the project was too high. Project staff and participants received friendly, and sometimes barbed, ribbing about the amount of media exposure the project received.

In fact it was difficult to strike a balance between having enough exposure to raise awareness and being overexposed. Jeremy Lewis commented that in order to gain a high profile at a national level, it was almost inevitable that the project would become a 'tall poppy' at the local level. By 1988 project staff no longer had to seek media exposure as the project received sufficient publicity without any self-promotion.

The project staff had trouble in convincing media people to focus attention on the whole farm planning process, rather than only on tree planting. It is much easier to write about 'kids on farms planting trees to fix salt', than it is to explain the need for rethinking farm layout and management for sustainable productivity. Several journalists, including Ian Doyle of 'Countrywide', Rod Usher of the Age and Nicol Taylor of ABC Radio 3WV, understood the whole farm planning process and portrayed it in a thoughtful, well-balanced manner. However on many occasions project staff were frustrated by the media obsession with trees. In an attempt to counter this, we wrote twenty-two articles ourselves, mainly for the local and rural press.

It was obvious by mid-1986 that progressive local farmers were very aware of the Potter project, but they, too, commonly associated it with trees rather than with whole farm planning. This intuitive assessment was backed up by the 1986 attitude study carried out by Cary, Beel and Hawkins. The study found that two-thirds of the forty-one randomly selected farmers interviewed in the Shire of Dundas were aware of the Potter project by the middle of 1986, compared with only one-fifth of forty-one randomly selected farmers interviewed in the Charlton Shire. However the Dundas farmers who were aware of the project identified it closely with trees, whereas the Potter Farmland Plan farmers emphasised whole farm planning and reversal of land degradation as being more important than trees. Awareness of the project among the Dundas farmers correlated significantly with the level of education attained. The more highly educated farmers were more likely to have heard of the project and it was interesting that the farmers who were aware of the project were more likely to be aware of land degradation on their own farms and to be in favour of tree planting.

While media exposure was useful for raising awareness of the project, the great 'selling point' of the Potter Farmland Plan is the impact it has on visitors who see the farmers and their work on the demonstration farms. The farm inspection tours and field days were by far the most successful means of promoting an understanding of whole farm planning.

In general, visitors' responses to the field days can be summarised as follows:

1 They were inspired by what they saw.
2 People were prepared to travel a long way (up to 500 kilometres) to attend.
3 Visitors consisted of approximately 50 per cent locals (less than 50 kilometres away) and fifty per cent non-locals.
4 Few people attended more than one field day out of the eleven held.
5 Nearly all people said that they would love to come and see the farms again in a few years' time.

The few negative comments about the field days were mainly about the amount of time spent in buses instead of walking over the farms or talking to the farmers. These difficulties were overcome for the 3000 or so people who were shown around the demonstration farms in smaller groups from 1985 to 1988. As mentioned in Chapter 7, these tours averaged less than twenty people per tour, which was a more manageable number, allowing more interaction between the tourists and the farmers and with the tour leader. These groups often came from other regions and later from other states. Initially the visitors to the project were mainly non-farmers with a professional interest in farm planning, land management or revegetation; however by 1988 farmers made up the great majority of visitors.

Project staff found it difficult to share farm inspection tours evenly between the demonstration farms and by the end of 1986 it was obvious that 'Helm View', 'Willandra' and, to a lesser extent, 'Wyola' and 'Daryn Rise' were toured most often. As well as having more extensive works and very committed farmers, 'Helm View' and 'Willandra' are closer to Hamilton than the other areas and many groups were pressed for time. Most groups came from east of Hamilton, which tended to deter them from Wando Vale, although those who did inspect Wando Vale demonstration farms invariably appreciated the opportunity. The visual impact of the project works is also highest on 'Helm View', where the stark landscape of 400 hectares of ringbarked trees reaching skywards in despair contrasts with the lush pasture and vigorous clumps and lines of young trees and shrubs developing beneath them. The aspect of 'Helm View', sloping gently from north-east to south-west, allows a view of the entire farm from one point, and the impact this view had on the visitors made it difficult for project staff to resist stopping there as the 'finale' of a trip around the demonstration farms.

The farm inspection tours and the involvement of local people in project works were the most successful public relations activities of the project from 1985 to 1988.

The public relations and extension activities of the Potter Farmland Plan occupied an increasing proportion of staff time as the project developed. For example, I conducted ninety farm tours and spoke about the project at eighty-three functions by the end of March 1988 and John Marriott conducted seventy-five farm tours and spoke at forty-six functions to the end of August 1988. The frequency of inquiries and demands for tours of demonstration farms increased steadily, from 1985 to 1988, to the point where, in March 1988, responding to these demands became a full-time job for one person. John Marriott absorbed some of this demand into his workload, assisted by his wife Sue, who was employed casually. It became clear that the demand for this service quickly increased to match the resources made available to meet it and that any future project should include a full-time 'tour guide'.

● ADOPTION OF WHOLE FARM PLANNING

The Potter Farmland Plan demonstrations should be examined from two separate perspectives. The primary objective of the project was to develop and demonstrate a *planning process*. The process by which the whole farm plans were developed is fundamental to the demonstration value of the project. The *application* of the process, the measures actually implemented on the demonstration farms, is the second component c i the demonstration. One of the difficulties with extension of the concepts of the Potter Farmland Plan is that the scale and impact of the works have led m: ·· people to confuse the whole farm planning process with its expression on the farms, hence the perception that 'Potter is about planting trees'.

There was an explosion of interest in whole farm planning in western Victoria from 1986 onwards, as the graph below illustrates.

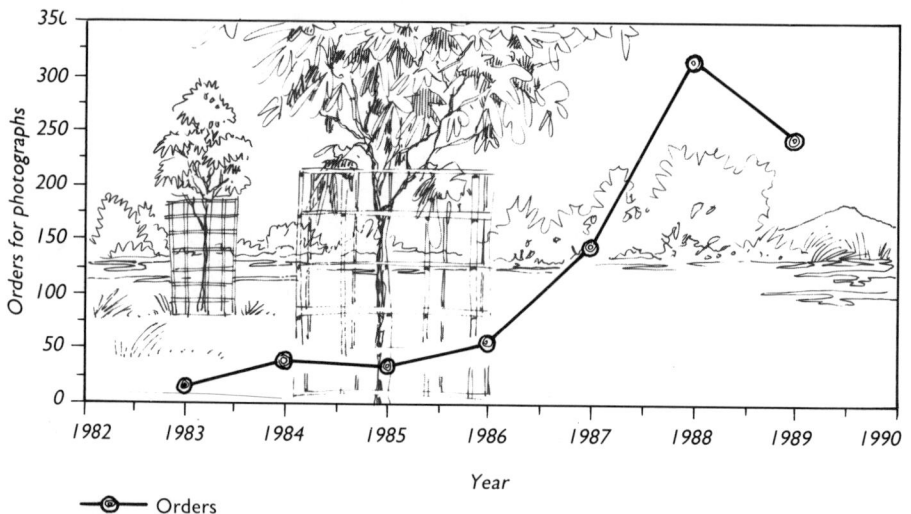

Orders

• Orders for aerial photograph base maps for farm planning, Hamilton office of the Department of Property and Services

The graph shows that farmers not only had an interest in farm planning, but that they were keen to use the same method of plan preparation (based on an aerial photo) as the demonstration farms. The accelerated interest of farmers in obtaining aerial photographs was not solely due to the Potter project. After a couple of short courses in farm planning, for which all participants were assisted to organise an aerial photograph, the Hamilton office of the Department of Property and Services had developed a simple system for ordering and paying for aerial photographs enlarged to a scale suitable for farm planning. When they were confident with their system they began to promote this service, particularly at 'Hamilton Sheepvention'. This helps to explain the big jump in orders for 1989.

• We lost few opportunities to hammer the principles and benefits of whole-farm planning: (above) Andrew Campbell and John Marriott at a Wando Vale Field Day in March 1986; (below) Bruce Milne explaining the 'Helm View' plan to members of the Wheat Research Council in October 1989.

• This montage of photographs from various field days hardly does justice to the effort put into getting the message about whole-farm planning across on every possible occasion.

The demand for short courses in whole farm planning was much higher in the Hamilton area than in other areas, and the demand for whole farm plans from private consultants also increased in the region.

Thus the local impact of the project on the adoption of whole farm planning was significant. The wider impacts are harder to estimate. There is no doubt that interest in farm and catchment planning and acceptance of the need for a whole farm approach has developed rapidly around Australia in the latter years of the 1980s, but it is difficult to pin down the extent of influence of the Potter Farmland Plan in this development. The key differences between the approach to farm planning we evolved at Hamilton and the previous approaches of soil conservation agencies and agricultural advisers are that the Potter approach emphasises that the farm planner is the farmer, not an outsider; that plans are only likely to be implemented if the land user is involved at every stage in the decision-making process; that the planning process is on-going and there is no such thing as a 'finished plan'; and that land classes, pasture, crop types and rotations, water and trees must all be considered together over the whole farm. These precepts are now much more commonly accepted and applied than they were in 1985.

I am aware of exciting whole farm and catchment-planning projects in a number of regions around Australia in 1990. Several of the Western Australian in 1990 Land Conservation District Committees were undertaking farm- and catchment-planning projects in which farmers were involved in preparing plans for their own properties, assisted by a consultant who ensured that the farm plans were based on a similar understanding of local conditions, land-degradation processes and sustainable land-management practices, so that they would be coherent when aggregated at the catchment level.

There has been a move to integrate Geographic Information System (GIS) technology with farm planning in Western Australia, encouraged by the Department of Agriculture, and similar moves are taking place in most other states. This technology, which was discussed in Chapter 3, enables data on geology, soils and vegetation, as well as man-made structures such as roads, fences, dams, drains, shelter belts and buildings, to be digitised and stored on computer. A big advantage of GIS, apart from the large amounts of information which can be stored, is that once information is in the computer, the time taken to draw successive drafts of any plan is minimised. Another advantage is that the scale of the plans can be manipulated without redrawing, which means that preparation of a catchment plan by aggregating farm plans is theoretically straightforward.

While I concede the advantages and usefulness of GIS to land-conservation and land-management agencies who have the necessary technical expertise and equipment to benefit from the technology, I remain concerned about the degree of farmer ownership which can be achieved if the farmer has to contact a public servant or consultant to change the plan. The beautiful computer-plotted plans tend to inhibit farmers from scribbling ideas or changes on them and tend not to be carried around in utes or nailed to shed walls like aerial photographs (particularly photopositives) can be. Until farmers or farmer groups have their own GIS equipment (as the Jerramungup Committee has done in Western Australia), I prefer to see the two formats working in tandem, with farmers constantly refining their plans on an aerial photo format, and government agencies maintaining digitised plans and smaller scale catchment plans, updated from time to time in consultation with farmers.

At the local level, the 'ripple effect' of other farmers adopting measures they saw on the demonstration farms was quite obvious by 1987. Several neighbours of Potter Farmland Plan farmers who had rarely planted trees before the project began to establish shelter belts and clumps using exactly the same techniques as the project farms. Sales of 'F41' weldmesh in Hamilton increased by a factor of ten after it was recommended for tree guards by project staff in articles in the *Hamilton Spectator*. The UV-treated plastic sleeves, which had not been available (due to lack of demand) in Hamilton before the project, became best-selling items—one Hamilton supplier alone was selling more than 50 000 guards per year by 1988. Sales of tree guards, stakes and trees increased sharply in the area between 1985 and 1988, although factors other than the project contributed to this increase, including rising wool prices, good seasons and a general increase in aware-ness of the value of trees on farms.

The Cary, Beel and Hawkins study was carried out in mid-1986, but there were already clear differences in attitudes between the demonstration farmers and the forty-one farmers randomly selected from the Dundas Shire. Demon-stration farmers identified themselves much more closely with increased farm productivity, improved management of the farm, whole farm planning and reversing land degradation than the Dundas farmers. Both groups were indifferent to government advisory services and the Dundas farmers identified themselves more closely with tax and cash incentives and shade and shelter than the demonstration farmers.

FACTORS INFLUENCING ADOPTION BY OTHER FARMERS

This section considers the factors influencing adoption of the Potter Farm-land Plan approach from two angles—the development of whole farm plans and their implementation.

Attitudes

The Potter Farmland Plan has demonstrated that the factor which has the most influence on the development and refinement of whole farm plans is the commitment of the individual farmer to the planning process. The most effective extension programme would enable individual farmers to see that the development of a whole farm plan was in their interests, rather than relying on government extension agencies telling farmers how to manage the land.

In order to create a climate of opinion in which farmers' enlightened self-interest stimulates adoption of whole farm planning, there need to be some fundamental shifts in attitude throughout the Australian community. It is not enough to recognise land-degradation problems and to want to fix them. Sustainable agricultural land management demands that production systems should complement the natural ecosystems upon which they depend. Bruce Milne's journey in thought, described at the end of Chapter 5, is an example of the changes in knowledge and belief that can be effected by direct experience, and of the qualitative changes needed to achieve the necessary redistribution of resources and effort *within* Australian farming systems.

Awareness campaigns have started the process of encouraging people to recognise that past and present land management has been inappropriate in many instances but, at this stage, the collective consciousness of the community seems to be directed towards correcting the mistakes of the past. Few seem to have taken the next step—to determine the sustainable capacity of the land and set production goals accordingly, rather than regarding maximum production as the ultimate goal. Optimising production to ensure the most efficient ratio of inputs to outputs within the constraints of sustainable land use should be the aim and the criterion for success. At present the gross level of production and the price received for the outputs are the yardsticks which measure the difference between 'good operators' and the rest. Extension activities, reflecting government policy, have en-couraged this attitude, and the winners of farm management competitions based on production gains over a certain time are accorded higher status than winners of the few conservation-oriented competitions such as those run by soil conservation agencies.

The images of the 'good farmer' and the 'good farm' need to change in the minds of farmers and their advisors—extension officers, bankers, accountants, stock agents, machinery and chemical salesmen, farmer unions and insurance agents. More importantly, these images need to change in the minds of the other 85 per cent of the population who live in the towns and cities. Farmers' attitudes are more likely to change in favour of sustainable land use if they are supported by a majority of public opinion. This was certainly the case in the early times of European settlement, when the puritan work ethic and the common perception of the 'good farm' as a neat chequer-board of cleared and ploughed fields encouraged farmers to battle against a harsh, unfamiliar landscape, attempting to mould it to their image of productive land. The old-timers' efforts were supported by the entire com-munity and it was consequently politically attractive for governments to offer tax concessions for clearing and grants for 'opening up' new land.

This may seem far removed from the adoption of whole farm planning, but part of its value is that the process itself encourages farmers to re-examine farm layout and management in the light of the capacity of the land, and their own needs for survival. The incentive to prepare a whole farm plan will be generated by the farmers' desire to farm the land in a more sustainable way. There are also incentives for preparing a whole farm plan even if the main goal is maximising production and profit, but the Potter Farmland Plan approach is based on taking the capability of each land unit into account and specifying farm improvements and management accordingly. Thus a desire for more sustainable land management is an essential prerequisite for whole farm planning, and the basis of this desire must be a change in attitude throughout the population.

I have yet to see a more striking example of the impact of attitude change on behaviour than Bruce Milne, who firmly believes that to change attitudes is the key to halting land degradation in Australia. During 1989, Bruce analysed the changes in his own attitudes, during the 1980s, which were greatly stimulated by involvement in the Potter Farmland Plan. He wrote down these changes and sent them to me. They illustrate a profound shift in thought and commitment, a shift which in Bruce's case is still occurring.

The table opposite outlines the changes as Bruce saw them.

Coordinated Advice

Two full-time staff working on fifteen farms for the first three years of the project, even with the implementation and public relations activities of the Potter Farmland Plan, is a much more intensive input of advice than the average farmer can expect from government extension services. Apart from the advantages of availability, we knew the demonstration areas intimately. Knowing paddock names, gate locations, where not to drive in winter, the cold and windy spots, the characteristics of different land types in different seasons, the places where sheep camp and the way in which the farm is managed is tremendously useful for an adviser assisting a farmer to prepare a plan, and gives the adviser more credibility.

Awareness of the Problem	
1980	*1989*
Global degradation is widespread, but my farm is not too bad.	Global degradation is widespread and life threatening. My farm is seriously degraded as part of the global problem.
The Western District Red Gum belt is beautiful.	Yes, but it has been radically changed by agriculture and is seriously ill as a result: • an aged monoculture with no chance of natural regeneration; • increasingly saline and waterlogged soils; • devastation of native wildlife due to habitat removal; • soils increasingly acidic, compacted and eroded by water, wind and stock movement.
The Australian farmer is the most efficient in the world.	There is a huge margin between the average Australian farmer and the best Australian farmers. Most farmers have plenty of room to improve production and viability.
I will need to squeeze as much as possible from my farm to survive the tough times ahead.	I must return my farm to good health, so that it can sustain production into the future and allow us both to survive.
Destruction of the Brazilian rainforest must stop.	Yes, but with our assistance, as the western world owes the Brazilians a land debt and an alternative lifestyle.
I wonder if all the effort required to produce wool and beef each year is worthwhile. I won't have anything lasting to leave my kids.	Implementing my farm plan is the most important and rewarding project I will ever be involved with. My kids and future generations will receive lasting benefits from my efforts.
The woodchip industry is damaging our forests.	The clearing of open woodlands for farming has been far more ecologically devastating than the forest harvest.

1980	1989
I know where I can plant some trees if I can get some time after drenching, shearing, ploughing…	I have prepared a whole farm plan. I know when and how to implement the plan which receives TOP PRIORITY.
With some government grants, I could fence out a couple of gullies and plant some trees.	I don't need any outside help to implement my farm plan over time.
Massive government assistance for on-ground works to halt land degradation is essential.	NO! Most government assistance should be directed towards education and awareness campaigns on land degradation problems and how to cure them. Minimal assistance and incentive schemes for on-ground works would be helpful in targeted areas of real need.
Planting trees on my farm is a big job. Fencing is time-consuming and costly. Planting with a spade is hard work, then I have to water them in. Often, survival rates are disappointing.	Planting trees on my farm is a breeze now. Electric fencing is effective, fast and cheap. The Hamilton Tree Planter makes planting faster, easier. Deep ripping and good weed control has eliminated watering. Direct seeding has great potential.
'Greenies' are radical irritants and a threat to good farmers.	I am a greenie.

Through their contact with project staff and exposure as the owners of demonstration farms, the Potter farmers also had access to advice from Department of Conservation, Forests and Lands staff on water supply and soil conservation issues, CSIRO researchers on dry-land salinity, Bill Middleton on tree species for wildlife habitat or DARA specialists for pasture establishment. These diverse sources of advice were integrated by project staff and the farmers themselves in the farm-planning process, which is one of its key virtues. A major difficulty for an individual farmer considering a whole farm plan is that extension services in most states are disparate and specialised. This is not necessarily invalid, as each person cannot be an expert in every aspect of farm management, but it tends to lead to territorial divisions of responsibility. As a consequence, it is difficult to obtain advice about the farm as a whole. If extension officers from different agencies and disciplines pooled their resources more and worked in teams, their advice could contribute to a whole farm approach.

An underlying factor which compounds the compartmentalisation problem is the lack of compulsory ecological training for agriculturalists. Agricultural training should be based on the necessity to see a farm as a functioning system, made up of interrelated and interdependent components superimposed over a natural ecosystem with its own internal dynamics. Gradual specialisation in training, a lack of interdisciplinary cooperation in research and extension and a lack of ecological understanding of the sustainable capacity of soils, vegetation and wildlife has led to the development of inappropriate farm-management systems. This is not a problem peculiar to Australia; it is becoming widely recognised throughout western industrialised countries.

Extension activity in Australia has historically been one-to-one advice from an extension officer responding to an individual farmer's need or problem. Apart from problem solving, most extension work has consisted of transferring research findings about how to improve production to the farming community—the goal of maximising productivity has been paramount. During the 1980s and early 1990s, funds to government extension services have been restricted and a search for more efficiency has led to renewed emphasis on group-based extension. Once again, gaps in training have been exposed as the limited amount of training that does exist in extension is mostly directed towards one-to-one programmes. The sociological skills of dealing with groups need to be developed within extension agencies' training courses. This need has been highlighted by the rapid growth in the number of landcare groups around Australia, many of which are engaged in whole farm planning.

It is important that farmers preparing whole farm plans receive consistent professional advice—consistent between its diverse sources, consistent with a whole farm approach and consistent with sustainable land management. The latter point raises a separate issue—that of the information base from which extension services and farmers operate. It is of little use attempting to plan the use of a certain land type within its capacity, if the capacity of that land type is not clearly understood. In many areas of Australia, information about land systems is inadequate.

However, the onus for gathering this information should not be on government agencies alone. Farmers can and should do a great deal more basic data collection and monitoring on their own farms. It is not enough to think of budgets simply in terms of dollars. If we are to develop sustainable agricultural systems we need to prepare soil budgets, water budgets, nutrient budgets, energy budgets, even biodiversity budgets to attempt to quantify

long-term impacts on the land. These analyses are complex to do scientifi-
cally, but they are well worth working through, if only to illustrate how little
we know about the dynamics and impacts of many land uses on many
land types. But there are rough sums that farmers can do, such as working
out the quantities of basic elements such as nitrogen, potassium and phos-
phorus removed in each tonne of grain and comparing this with the amount
replaced in fertilisers or from other sources. It is salutary for cropping farmers
in areas with potential dry-land salinity problems, such as in south-west
Western Australia, the Eyre Peninsula and north-central and north-west
Victoria, to do a basic water budget comparing rainfall with run off, infiltration and evapotranspiration to get an idea of the extent to which current
cropping systems 'leak', thus causing water tables to rise. It would be desirable
to have a situation where farmers are demanding this information, and doing
everything within their capabilities to obtain it.

Implementation of Whole Farm Plans

Attitudes

The attitudes which cause farmers to change their priorities in favour of
works which protect their land in the long term rather than exploiting it
in the short term are graphically illustrated by the changes in awareness
described by Bruce Milne earlier in this chapter. While it is naïve to suggest
that the ability to implement a whole farm plan is not influenced by the
amount of time, labour and money available, the farmers' own priorities
determine how those resources are allocated.

It is clear from the Potter farms that when farmers are confident that
they can do a job successfully, whether it be to design and put in a laneway,
sow a new pasture or plant some trees, then they are much more likely
to do it. The 'can-do' attitude is very important and it is closely linked to
the quality of advice and local demonstrations.

Financial incentives

The works carried out on the demonstration farms cost an average of $145
per hectare, including the farmers' labour costed at contract rates. The whole
farm plans were implemented at least three times faster than would be
expected of a farmer working without help. Thus, if the project works were
costed over ten years rather then three, the treatment cost would be about
$14.50 a hectare, including labour, each year. Works aimed at redressing
land degradation are tax deductible under section 75(D) of the Income Tax
Act, which means that the actual cash outlay for the average farmer is probably

about $5 a hectare each year. As the average annual net return per hectare of the forty-four farms in the south-west monitor farm project, from 1987 to 1988, was $140, an extra $14.50 spent on sustaining the productive capacity of the basic resource is easy to justify.

Nevertheless there are financial deterrents which, if removed, would make it more attractive for farmers to do works which do not yield a short-term cash flow.

For example, it would help if land taken out of production for tree establishment was recognised either in the tax or rating systems. In Victoria the rating system, based on improved capital value, can penalise farmers who improve the value of their land with trees. Some incentive within the rating system could be provided by deducting the area taken out of production for conservation purposes from the rateable area. Municipalities no doubt argue that to maintain their rate revenue they would have to increase rates from other areas, so perhaps some federal assistance through the taxation system would be more appropriate.

● CONSULTATIONS

The consultation at Creswick which began the project was briefly described in Chapter 2. Several other consultations were organised and run by Peter Mathews during the course of the Potter Farmland Plan. Each of these followed the general model of open discussion between twenty to forty participants for two days, with an independent chairman, no media and no formal resolutions.

The next consultation after Creswick was held at Monash in April 1985. Representatives of relevant Federal and State Government departments, research and training institutions, the media, agribusiness, the law, and two farmers, Bruce Milne and Bill Speirs, examined the general changes needed to help farmers get on with the task of improving land management.

Further consultations about land management issues and the potential contribution of whole farm planning were held in Ballarat in 1986 and the Tasmanian Midlands in June 1987. The Tasmanian consultation was particularly successful because it led to a whole farm planning project based on three demonstration farms in the Midlands, organised by Greening Australia, with significant local private sponsorship and assistance from the Tasmanian Department of Primary Industries and Forestry Commission and the National Soil Conservation Program.

● HAMILTON REGION 2000

The Hamilton Region 2000 project was established to complement the Potter Farmland Plan as a model for Australian rural development. The Potter Farmland Plan concentrated on demonstrating what individual farmers can do with their own land, whereas Hamilton Region 2000 was designed to assist an entire rural community, including country towns, to plan its future development, taking environmental, social and economic factors into account.

The approach adopted was significant because it raised the issue of the balance between 'grass-roots' development, or working from the top down. The philosophy underlying Hamilton Region 2000 was that a facilitating process can assist people to become aware of changes in the world around them and to change their attitudes and practices accordingly, but that the pace of these processes is dictated by the people involved, not by the facilitators.

Hamilton's population has remained at 10 000 people for the past forty years, but most of the surrounding municipalities have suffered steady decline in population.

The Hamilton Region 2000 project complemented the Potter Farmland Plan, since the logic of planning and managing the land, blending ecology and agriculture to maintain and enhance productivity, dovetailed with planning and managing the region in harmony with its social, environmental and economic base. Hamilton Region 2000 aimed to stimulate the people of the region to develop community plans based on opportunities for achieving community goals and enhanced economic possibilities.

In June and August 1986, the Potter Farmland Plan ran thirteen small discussion groups and then a follow-up consultation in Hamilton. The issues discussed included community understanding of land degradation, the reaction by the community to 'greenies' and ways to increase communication between city and rural people, especially about land degradation and the wider community's responsibility to support the efforts of those trying to tackle it. The primary economic base of the region is its land. During the consultations it was suggested that a project to raise awareness in the wider community of how to adapt to change was needed, rather than acquiescing in the general trend of decline for country towns.

Funds from the Dame Elisabeth Murdoch Trust enabled a project to be started in early 1987. Peter Mathews and John Jack of the Potter Farmland Plan Executive were to oversee the project for two years. Victoria Mack, a local farmer with wide community interests, was appointed Community Facilitator in April 1987. An advisory group was also established to support the project, comprising a cross-section of respected and interested citizens.

The project grew very quickly to embrace economic and social development, value-added enterprise, consultation and community discussion, education issues and tourism. The Board of Hamilton Region 2000 aimed to establish more local manufacturing, and new enterprises in the region. To achieve this aim, the project did not focus on attracting industry to the town to provide employment. It took the view that sustainable industry must be locally run and use local resources, in conjunction with outside expertise when required and access to working capital. 'Thinking globally and acting locally' was the principle of its activities.

There are inherent problems in this approach, as Hamilton Region 2000's staff and advisory board discovered. New or large-scale economic activity is often beyond the resources and experience of country people and small country firms. The region needs resource people to be sympathetic and able to help people to find the right information and support. Local people asked, at the beginning of the project, why they should develop at all. They often felt quite comfortable as they were, either oblivious to or unaffected by any decline. A common attitude was, 'The Hamilton Region is very nice the way it is, and we don't want our environment polluted by industry or our community troubled by rapid social or economic change'.

Initial awareness-raising activities included an essay competition for schoolchildren entitled 'Hamilton in the Year 2000 AD' and 'Farm Discovery Tours' to the Potter farms during winter, 1987. These activities were followed in October 1987 by a national conference bringing the decision makers and heads of State and Federal structures together with the Hamilton community to consider at first hand the reality of the decisions which affect rural development. The three-day conference was followed by a public meeting in Hamilton, attended by more than 450 people.

Following the conference and public meeting over 200 people took part in twenty-seven community discussion groups, which raised interesting ideas for future development as well as concerns about Hamilton's conservative nature, fear of change and social divisions within the region. There was agreement that the region could flourish if these problems were overcome. Many of the local people brought together in these simple discussion groups found it an enjoyable learning experience and many acknowledged it as the first time they had ever really considered the issues raised.

Following the release of the discussion group reports in February 1988, two 'action workshops' were held in which interested people were invited to participate in small groups working on four main issues identified in the report: value-added enterprise, tourism, community development and education.

All these workshops resulted in continuing activity of various kinds, such as:

- A proposal for a National Wool and Rural Industries Skills Training Centre in Hamilton, from the education group. The idea was based on two assumptions; that the Hamilton Region has a national reputation for its production of fine and superfine wool and its management of high quality pastures and animal husbandry techniques and that there was an urgent need to improve the training and retraining opportunities for people in the pastoral industry. The WRIST Centre, as it is now known, will work closely with the South-West College of TAFE, the Victorian College of Agriculture and Horticulture, Melbourne University and La Trobe University. It received $35 000 in 1989 from the Federal Department of Employment, Education and Training, the Victorian Education Foundation and the State Training Board of Victoria, to establish a business plan, management structure and needs analysis for the proposal.

- Establishment of development committees at Balmoral and Dunkeld (smaller towns north and east of Hamilton respectively) which have attempted to do on a local level some of the awareness-raising and catalytic activities of Hamilton Region 2000. Local projects have included production of a Balmoral tourist brochure and community directory and a Dunkeld townscape-development plan and art shows and community craft markets. The Dunkeld group has also proposed to plant an arboretum in the heart of the township. The development of stage two of the WRIST Centre was provisionally approved in November 1990, with funding to be provided by The State Training Board. The Centre will be operational by 1992. The Commonwealth Government has also indicated strong support for the Centre's development, under The National Skills Training Centre.

- Community facilitator training. During 1988 Peter Mathews and John Jack received a grant of $45 000 from the Local Government Development Programme to fund the training of twelve volunteer community facilitators and to conduct follow-up community consultations on social and economic development and education. The course ran for eleven days over six months. Dr John Bailey from the Chisholm Institute of Technology and Peter Mathews were the main instructors. The course aimed to train the facilitators to help their own communities to adjust to rapid social change.

- An economic development consultation held in February 1989 led to the formation of the development company Wool Capital 2000 Pty Ltd. The board of directors of Wool Capital 2000 comprises local business people and meets each Friday morning. New projects have been developed and existing enterprises supported or expanded. New projects have included the development stages of a flower farm, Cassette Tours of Australia and investigation into a meat-rendering plant. Supported projects have included Australian Sheepskin Products Pty Ltd, Gro-Beta Organic sheep manure and Ultrafine Pty Ltd, manufacturers of ultrafine woollen knitwear. The company has sixteen shareholders and ten directors and raised $8000 in share capital to get started.

- Hamilton Region 2000, together with Hamilton Adult Education, organised a training course for tour guides with twelve interested people during 1988. Six of the participants formed the Hamilton District Tour Guides early in 1989, and their activities have expanded significantly during the year. The growth in their personal confidence is a fine example of what the Hamilton Region 2000 project can achieve. Hamilton Region 2000 also supported the development of Hamilton Environmental Awareness and Learning (HEAL), discussed later.

- Enterprise development and support. Hamilton Region 2000 has answered requests for information and contacts and helped to develop ideas from many people in a diverse range of enterprises in the region. For example, the establishment of Hamilton Community Printing and Typing Service and Hamilton Laser Colour Copying in March 1989 was supported by the project. Hamilton Region 2000 assisted Australian Sheepskin Products to raise local equity investment in the company and to prepare a company profile. Another contribution from the project was its intervention in a dispute with the Health Department about registration of a special accommodation house. The project helped the client to prepare a case for the development and, together with a local politician, the client's lawyer and the client's architect, persuaded the department to reconsider its decision. The development is to commence early in 1990. The City of Hamilton recognises the contribution of Hamilton Region 2000 in the area and provides $10 000 to the project each year.

In 1990 Hamilton Region 2000 became a managing agent for the Commonwealth, to deliver the New Enterprise Incentive Scheme (NEIS) in the region. NEIS provides training in business plan development to unemployed people with viable ideas for self-employment and small business.

Hamilton Region 2000 has attempted to cover all kinds of development—economic, social, environmental and educational. There are often problems and snags. For example, a consultant's report or feasibility study may be inconclusive or it may recommend a second report, people change their minds on directions and others never come back after initial investigations, projects often take much longer than expected and current interest rates are not encouraging new or expanded enterprise investment.

Hamilton Region 2000 has demonstrated that people with ideas, commitment and the desire for personal or commercial achievement make things happen. The best will in the world and the highest degree of altruism by community members will not necessarily get things off the ground unless individuals are prepared to commit themselves to on-going effort. The project has shown that it takes a great deal of persistence and energy to help communities to understand the benefits of working together and planning for a future which they have chosen. It has complemented the demonstration farms of the Potter Farmland Plan in showing how a rural community as a whole can start to improve its environmental, social and economic circumstances.

Late in 1990 Hamilton Region 2000 was asked to provide a Secretariat to over twenty district woolgrowers, to help them take positive action in the face of the rural crisis bearing down on the region in 1991.

● THE CENTRE FOR FARM PLANNING AND LAND MANAGEMENT

During 1986 and 1987, towards the end of the first phase of the Potter Farmland Plan, the Executive began to concentrate on more general changes needed to complement the efforts of the individual demonstration farmers, in order to develop more sustainable systems of land management. The first of these was Hamilton Region 2000.

The second of these developments was the Centre for Farm Planning and Land Management at the University of Melbourne, which was established in 1988 with funding for three years from the Elisabeth Murdoch Trust, the Sidney Myer Foundation and the Rowden White Bequest. The aim of the Centre is to conduct and encourage research into sustainable land use systems and to facilitate interaction between the people and institutions involved. The Centre is intended to be multidisciplinary and to act as a catalyst, blending ecology with sociology to pull together disparate areas of research and promote a more holistic approach to agricultural research and land management.

The establishment of the Centre was a response to the criticism of scientists visiting the demonstration farms, who often asked how we would know which elements of the whole farm plans contributed how much, or how many replications of different treatments we were doing, or what we were using as a control. These are valid questions for a research project, but not for a project trying to demonstrate an integrated whole farm approach so that farmers can see how it fits together and assess its effectiveness themselves, just as they have to integrate information every day in making decisions about their farms. The Centre for Farm Planning and Land Management is within a traditional university and its advisory board consists of professorial staff and representatives of funding bodies, government, farmers and John Jack, Peter Mathews and Pat Feilman from the Potter Farmland Plan. It attempts to demonstrate how researchers and academics may integrate their disciplines to provide more appropriate answers for land users.

● POTTER FARMLAND PLAN LTD

The third main achievement of the Potter Farmland Plan Executive after the demonstration activities at Hamilton was the development of a TAFE-accredited course in Whole Farm Planning at the Victorian College of Agriculture and Horticulture, Glenormiston. This was assisted by a grant from the Victorian Education Foundation and whole farm planning is now institutionalised within the adult education system. The Whole Farm Planning course is complemented by the two-hour series of videos mentioned in Chapter 7.

The Potter Farmland Plan Executive was officially dissolved in June 1988 and replaced by a nonprofit company called Potter Farmland Plan Ltd. This company was formed to coordinate the continuing documentation of the project and the development of the whole farm planning short courses and videos, and to promote whole farm planning throughout Victoria and in other states. The board of Potter Farmland Plan Ltd comprises John Jack, Pat Feilman, Peter Mathews, Bill Middleton, Bill Speirs, Bruce Milne and Geoff Handbury, a farmer from Kanagulk who also represents the Elisabeth Murdoch Trust.

● HAMILTON ENVIRONMENTAL AWARENESS AND LEARNING

John Marriott's employment with the Potter Farmland Plan finished late in 1988. He immediately formed a business called Farm Planning Services, consulting to individuals and groups of farmers, running short farm planning courses. For the first time since the project began, late in 1984, it was without full-time staff at Hamilton, yet by this time showing visitors around had become a full-time job.

CFL agreed to appoint a whole farm planning advisory officer at Hamilton, but there were delays in getting this position established, and it was not filled until August 1989, when Peter Dixon was appointed to work half-time until October 1989, at which time he became full-time. (Departmental support proved to be short-lived, however; funding for the position was withdrawn in August 1990.) This gap in staffing was bad for the project's public relations —many prospective visitors to the project were unable to have the guided tours around the farms that others had enjoyed before 1989. As an interim measure, John Marriott was paid a contract rate by CFL to show people around, a task which he had to fit in with his private commitments.

Thus, for most of 1989, the Potter Farmland Plan was without effective extension support—certainly by comparison to the resources it had enjoyed from 1985 to 1988. Nevertheless, interest in the project and pressure from visitors remained high, and several of the demonstration farmers, notably Bruce Milne and Peter Waldron, carried a disproportionate amount of the load. The adjustment from a period of generous funding with permanent staff to a lack of funding and staff was difficult, particularly as outsiders continued to want to see and learn from the demonstration farms, which was, after all, the object of the project.

In hindsight however, this period was beneficial because it forced the people at Hamilton to assume complete responsibility for the project, rather than relying on the Melbourne-based Executive for direction. The pressure from farmers, scientists and community groups from all over Australia wishing to see the demonstration farms reached a point in 1989 at which the demand could simply not be satisfied with existing resources. In response to this demand and to a desire to see the maximum value realised from the effort which had been put into the demonstration farms, Sue Marriott gathered together a group of interested people, including several Potter farmers, staff from DARA and the DCE, a senior schoolteacher and chairperson of Hamilton Institute of Rural Learning, farmers from the Dundas-Black Range corridor group, and Victoria Mack of Hamilton Region 2000. Their aim was to ensure that people could still visit the demonstration sites—in fact they wanted to improve the access and information available to visitors and students, but in such a way that they could continue to do the farming, research or teaching which attracted the visitors in the first place. They named the new project Hamilton Environmental Awareness and Learning (HEAL).

After several meetings in July and August 1989, an outline and a submission for funding was presented to Greening Australia, the Victorian Tourism Commission, the Federal Department of Employment, Education and Training and the Victorian Education Department.

The key points of this submission were:

● The Hamilton region has a unique resource in its practical farm demonstrations, research trials, examples of coordinated whole farm and area plans, farmer commitment and farmer group activity. It has experts in whole farm planning, revegetation techniques and pasture establishment. Revegetation works for land degradation control, farm timber and shelter, agroforestry and direct seeding, and innovative, cost-effective techniques for tree planting and protection are the most advanced in Australia and their demonstration value improves each year.

- There is an urgent need for this combination of technical knowledge and practical demonstration to be accessible to interested people throughout Australia. Demand from visitors is increasing rapidly, without any advertising except by word of mouth since 1987.

- The demand for access cannot currently be met, yet the demonstrations deserve support, promotion, coordination of visits and integration into primary, secondary and adult education programmes.

As a result, Greening Australia agreed to provide HEAL with $35 000 for the first year from September 1989 and, at the time of writing, has committed a total of $55 000 to the project in a far-sighted sponsorship.

Greening Australia is a community-based federation of state organisations. Commonwealth funding of about $4.7 million per year for the One Billion Trees programme makes the government Greening Australia's largest sponsor, but not the only one, as many private companies sponsor individual projects. The overall aim of the One Billion Trees programme is simply to establish 1 billion trees in Australia by the year 2000, roughly two-thirds of which are to be established through direct seeding or natural regeneration. While $4.7 million represents a large funding increase for Greening Australia, it is only $0.047 per tree, which means that the One Billion Trees funding has to be catalytic, multiplying itself at least twenty times on the ground.

Greening Australia is well aware of this, and aims to encourage the right tree in the right place for the right reason, which is why the HEAL project is so complementary to its activities. Physically growing 1 billion trees in Australia is easy, but paying for their establishment and protection is not. Assuming a combination of clump, strip and wood-lot revegetation with an average density of 1000 trees per hectare, at least one million kilometres of fencing will be required for the billion trees, at a cost of roughly *two billion dollars*. This is a further reason why the demonstrations of efficient tree establishment and protection techniques in the Hamilton area are highly valued by Greening Australia.

Sue Marriott is the half-time coordinator of HEAL, organising tours for groups according to their interests and their needs for technical advice. Visitors pay only for the time contributed by the farmers on their own land, whilst other costs are subsidised by Greening Australia.

The aims of Hamilton Environmental Awareness and Learning are:

1 To promote sustainable land management, in particular the practical demonstrations on farms and research trials of the Hamilton region locally and throughout Australia.
2 To encourage and host visits to these demonstrations, providing technical interpretations, direct contact with farmers and a range of tour options to suit the requirements of different groups and to hand out documentation of results and educational material.
3 To interact with local and distant schools to provide 'hands-on' learning opportunities about rural environmental issues for children at primary and secondary schools and to support adult education programmes by providing access to practical demonstrations and technical expertise.

The Potter Farmland Plan is not the only attraction for visitors to the Hamilton area who are interested in sustainable agriculture. Whilst trees are a symbol of new approaches to farming, the Hamilton area has much more to offer. The strength of the demonstrations in this area lies in the integration of conservation with production, of pasture, stock management and farm design, providing an appropriate context for revegetation. For people interested in improving the ecological stability of farmland while maintaining or improving productivity, the Potter Farmland Plan has become an inspiration, but the project is complemented by some other excellent educational assets, described below.

AGROFORESTRY AND DIRECT SEEDING RESEARCH AND DEMONSTRATION SITES

Since 1984, Rod Bird and Keith Cumming of the DARA Pastoral Research Institute at Hamilton have established fifteen agroforestry trials (with 25 000 trees) using forty native and introduced species at various spacings on a range of soil types. Agroforestry refers to the integration of agriculture with forestry on the same site, where economic benefit is gained both from trees and from stock or crops grown among the trees. Agroforestry is in its infancy in Australia and, while its potential to improve productivity while tackling problems such as rising water tables has long been recognised, there is still a dearth of well-documented working examples in the paddocks. The trials around Hamilton are among the largest and most advanced in Australia, especially for species other than *Pinus radiata*. Most of Rod and Keith's demonstration sites are in eight-hectare blocks, which gives farmers a realistic impression of what a change in land use to agroforestry means in practice.

Rod Bird and Keith Cumming also received a National Soil Conservation Program grant to establish twelve direct-seeding trials on different soil types and climatic zones from the southern Wimmera, to the lateritic tablelands and the basalt plains, using twenty different treatments, over three years starting in 1987. Some of the treatments have achieved startling results and have added significantly to our knowledge of pre- and post-seeding herbicide treatment and soil disturbance. These trials have nearly all been established on real farms in open pasture, several of them on or close to the Potter demonstration farms. The Pastoral Research Institute also has a long-term shelter belt trial, which began in 1980. By late 1989, 95 000 trees had been planted, including 20 000 salt-tolerant trees, using various establishment techniques, to trial rehabilitation of saline discharge zones. Keith Cumming's Hamilton Tree Planter (described in Chapter 4) has revolutionised hand-planting techniques and is rapidly gaining a national reputation.

DRY-LAND SALINITY RESEARCH

The DCE has significant 'best-bet' research trials, examining the long-term effects of tackling dry-land salting by using perennial pastures and trees on subcatchments at Konongwootong (between Melville Forest and Wando Vale), Gatum (8 kilometres north-east of 'Helm View') and on 'Daryn Rise' at Glenthompson. One of the assumptions at the beginning of the Potter Farmland Plan was that there is enough existing knowledge of the fundamental causes of land degradation to enable farmers to act without waiting for more research. In fact farmers cannot afford to wait for research, they need to begin now to develop farming systems which use more water, systems which will be more productive almost by definition. It is equally important, however, that research continues, in order to scan and develop new options and to put figures on some of the 'best-bet' treatments. It is even better if research is carried out close to whole farm demonstrations.

The three subcatchments in the DCE research sites have been monitored closely to examine the impact of various treatments on water-table levels, run off and farm production, and this information complements the impact of the demonstration farms.

THE DUNDAS-BLACK RANGE CORRIDOR PROJECT

Six landholders living between the remnant forests of the Mt Dundas Range and the Black Range (western outposts of the Grampians), have coordinated their whole farm plans in order to create a network of indigenous vegetation.

They have linked up shelter belts, linear reserves on unused government roads and railway lines and revegetation on salt-affected drainage lines. Since forming a landcare group in 1987 they have planted 30 000 trees and direct seeded approximately 20 hectares of indigenous trees and perennial pastures on their farms in order to tackle dry-land salinity, create farm shelter and, importantly, to link the two forests with a network of wildlife corridors. The project was a finalist in the Greening Australia National Tree Care Awards in 1990, and featured in an inspiring presentation programme on ABC TV's 'Countrywide'.

The Dundas-Black Range group is one of ten or so landcare groups in the region. There are more than seventy groups in Victoria and six hundred in Australia, involving more than 15 000 farmers in late 1990. Landcare is a grass-roots revolution which is providing farmers with opportunities to get together and talk, learn, plan, find support for and take initiative in tackling their local long-term land conservation problems. Landcare is changing the relationship between farmers and land conservation agencies, and has the potential to have an enormous impact on Australian land use throughout the 1990s, particularly if the bipartisan support of political parties and groups as diverse as the Australian Conservation Foundation and the National Farmers Federation is sustained.

The group in the Hamilton area are already actively tackling problems. For example, the Wando Vale catchment group, which has built on the work of the Potter farmers and other leading local conservationist farmers, has gained funds from the National Soil Conservation Program to engage Rural Planning Victoria to help them prepare a catchment plan to establish priorities for land degradation control. Other groups at Mooralla and around Glenthompson have also been active and have experience relevant to other farmers in Australia involved in landcare.

THE PETER FRANCIS ARBORETUM

The Peter Francis Arboretum at the Points Reserve in Coleraine contains the largest range of eucalypt species on a single site in the world. The late Peter Francis, then an employee of the Shire of Wannon, began the Arboretum in 1966 on a 1–hectare site. It now covers 33 hectares and has many thousands of trees, including 400 species of eucalyptus and many species of acacia, banksia, casuarina, hakea and callistemon, all named and in their natural associations. The reserve has a part-time ranger (funded by Greening Australia and the Victorian government) and its own nursery and is of great value for all lovers and students of Australian flora.

THE GLENELG FARM TREE GROUP AND PRIVATE CONSULTANTS

One of the first farm tree groups in Victoria, formed in 1980, was the Glenelg group based at Hamilton. It includes some of Australia's leading farm tree growers, whose farms show what can be achieved by farmers with no external help except good advice from group activities. The Hamilton area is also home of two leading whole farm planning firms. John Marriott runs short courses in whole farm planning and prepares individual farm plans. Rural Planning Australia organises revegetation and consults in farm ecology, rural landscaping, and farm and catchment planning.

THE BANDICOOT RESCUE PROGRAMME

Hamilton contains the last remaining mainland population of the Eastern Barred Bandicoot, *Parameles gunnii*, within the town boundaries. The fate of this endearing creature is precariously balanced. The population of fewer than three hundred individuals is in danger of extinction because of predation by the thousand or so domestic and countless feral cats on the loose in Hamilton every night, road deaths, destruction of their habitat, agricultural and domestic pesticides and parasites. If the species manages to survive, it will be due to the innovative efforts of a few committed people who have stimulated a community programme of zoological and genetic research, community education, predator control and habitat preservation and re-creation, with government, philanthropic and corporate support. The Hamilton Institute of Rural Learning has revegetated and fenced out 100 hectares of community parkland as a bandicoot habitat which includes native grasslands of significant conservation value. The Institute has very valuable environmental education resources and a pivotal role in the bandicoot rescue.

Hamilton Environmental Awareness and Learning is a small organisation representing farmers, relevant government departments, private consultants, community groups and education interests. HEAL uses people and facilities already in the community to promote and organise regular visits to the area; provide tour guides, transport and technical information; and to contribute to environmental education programmes for school children throughout Victoria.

HEAL has two long-range visions; the first is to advertise a timetable for tours (perhaps one half-day tour and one full-day tour a week) in farming and environmental magazines and newspapers and among farm tree and landcare groups, government agencies and schools. Several different tours will be arranged, to suit the preferences of different groups, to allow for variable weather and to spread the impact of tours. The system will be flexible enough to cater for groups who have a particular requirement or a specific date, or who want to take part in activities such as revegetation work, surveys of farm wildlife, streams or water tables, or everyday farming activities such as shearing and haymaking. HEAL hopes that around 2000 people per year will continue to visit these farms and other environmental assets of the region.

The second long-term plan is to have a full-time seconded teacher from the Ministry of Education to involve primary and secondary students directly in rural environmental education and to liaise with adult education groups. At the time of writing we are still negotiating about this part of the project with the Ministry of Education, but the farm tours are working very well.

● CONCLUSION

Life after Potter for most of the demonstration farmers will never be quite the same as before. Involvement in the project for at least half the farmers meant that they were exposed to new ideas, new information, extra work and the scrutiny of a diverse range of visitors. Demonstration farmers often gained as much knowledge from the questions and reactions of visitors as they imparted. This process is continuing, thanks to the activities of Sue Marriott and HEAL, supported by Greening Australia.

The process of refining and implementing whole farm plans is continuing on most of the farms. The extent to which participants in the project have continued on with works without financial assistance is roughly proportional to their inputs during the project. In other words, those farms with higher levels of activity and farmer inputs during the project have continued at a higher level of activity since 1988 although, as was specified in the whole farm plans, the emphasis has shifted in some cases from long-term revegetation work to pasture improvements which will have a faster economic return.

This trend will probably accelerate for a couple of years with the downturn in wool and the general economy in 1990. However a point which should be evident by now is that whole farm planning is cheap. In fact, during economic hard times there should be even more incentive to plan—failing to do so is invariably expensive.

An outstanding feature of the Potter demonstration farms is that their visual impact is more striking each year as trees grow, saline areas are reclaimed with pastures and farm layouts are continually refined. I gain tremendous satisfaction from occasional visits to the Potter farms on trips home to my family farm and from anticipating the evolution of these farms over the next fifty years.

My secondment from the then Department of Conservation, Forests and Lands to the Potter Farmland Plan finished in March 1988, by which stage the first draft of this manuscript was well advanced. After a great deal of soul searching, I applied successfully for the position of Assistant Director of the Centre for Farm Planning and Land Management. I was keen to develop the concept of whole farm planning further, to play an active role in distinguishing clearly between the principles and the application of the process and to translate the benefits of whole farm planning into financial terms.

The Centre evolved differently, however, moving into the broader realms of sustainable agriculture. This provided opportunities to examine wider land-use issues from ecological and sociological perspectives and to meet policy and decision makers, interest groups, farmers and scientists. In this role I met Phillip Toyne, Director of the Australian Conservation Foundation (ACF) and Rick Farley, Executive Director of the National Farmers Federation (NFF), at the same time as these two organisations were forming the Ministerial Task Force on Land Degradation with the then Minister for Resources, Senator Peter Cook. I attended meetings involving the ACF and NFF which led to the development of a joint proposal for a National Land Management Programme, based on local landcare groups and a system of farm and district planning. This proposal was presented to the Prime Minister in February 1989 and became the basis of a major environment statement in Wentworth in July 1989. The Minister later asked me to become 'National Landcare Facilitator', a consultant who was to provide the national Soil Conservation Advisory Committee with an overview of landcare activities and report on the effectiveness of landcare and how the programme could be improved.

This role put me back in contact with farmers all over Australia, searching for ways in which the system can provide more appropriate help to those who are trying to develop better farming practices. It has also enabled me to see whole farm planning and the Potter Farmland Plan experience in a broader context. I see various methods of self-help or farmer-driven farm planning currently developing around Australia at first hand, in a range of different physical and administrative environments. I remain very comfortable with the principles and assumptions upon which the Potter Farmland Plan was based and with the farm planning process which evolved.

For the benefit of anyone considering a similar project, there are a few things I would do differently, if given another opportunity.

- Many people who talk to me about the Potter project still seem to get distracted by the number of dollars spent implementing plans, summing up the project as 'impressive, but too expensive for the average farmer'. This completely misses the distinction between the *planning process*, and the *implementation* of the plans, which was deliberately carried out three times faster than most farmers could afford. Perhaps we could have emphasised the planning process more and implemented fewer works, but I feel that demonstrations need to make a strong visual impact. Rather, I would reduce the number of demonstration farms, implementing the farm plans over a longer period while making sure that their impact on the farms chosen is profound.

- The local community—including neighbours and other farmers, local and state government agencies and small business—need to feel that they have a stake in the project from the start. We made a promising start with the Creswick consultation and the Hamilton 2000 project showed what is possible with integrated community development several years later, but the pressures of a hectic works programme and a constant stream of visitors prevented more active consultation with and involvement of outside groups in the project from 1985 to 1987. In retrospect, this is a pity. Greater involvement of townspeople, community groups and government staff throughout the project may have ensured an active, on-going monitoring programme and greater support for the demonstration farmers after the initial funding ceased. HEAL is now taking up the slack in facilitating visits, but the period of uncertainty immediately after Potter funding ceased (as we had known it would for three years), and the lack of a comprehensive monitoring programme were unfortunate.

- It is important to completely open up land users' minds to new possibilities, new ideas and new ways of thinking very early in the farm planning process. Getting the farmers together more often to seek information, identify their assets and opportunities and get a better understanding of their problems would have helped some participants to take control of the planning process for their farms earlier, ensuring a greater degree of ownership of the result. This is easy to say now that we have a better understanding of the farm planning process and now that farmers are planning for land conservation in groups all over Australia. But for the first two years of the project the whole farm planning process was still evolving. It is also important to remember that we set out to demonstrate what the individual can do on his/her own land. Group interaction is extremely valuable, particularly in farm and catchment planning, but group activity for its own sake can be overdone, particularly if group members do not live within the same 'social catchments'.

Hindsight provides a marvellous vantage point from which to say 'if only'. At the time we were in uncharted territory, subject to the constraints of time, the priorities and time frame of the funding body and the need to make an impact on the core of the project—the demonstration farms. The contributions the project has made in the areas of developing and demonstrating more holistic farm planning processes, more effective and efficient tree establishment and protection techniques, more effective shelter strategies and more efficient fencing have been significant, if difficult to measure.

The importance of farm and catchment planning in tackling land degradation and developing more sustainable land management practices is now widely recognised by extension agents, agricultural educators, landcare groups, government agencies and funding bodies throughout Australia. Farm and catchment plans are valued because they force a systematic appraisal of problems and solutions, hopefully resulting in a smarter allocation of resources; they provide a framework of integrating advice and information from various disciplines; and, if developed properly, they should stimulate approaches which integrate agronomic with ecological and economic solutions. At the very least, the Potter Farmland Plan raised the profile of farm planning in Australia in the mid-1980s, questioned traditional thinking and showed that farmers themselves are the right people to drive the farm planning process. The project did not just talk or publish brochures and papers. Ideas were put into practice on a large scale on *real* farms, and the story was told and is still being told by *real* farmers, from personal experience. This alone was worth the effort.

It seems many years since I started this book. Community concern for the environment; the debate about 'ecologically sustainable development'; a national landcare movement assisted by a coalition between farmers and conservationists, involving about fifteen thousand farmers in six hundred groups; a great expansion in rural revegetation activity and support; growth in awareness and acceptance of the potential of organic agriculture; widespread interest in new approaches to farm and catchment planning— all these developments have made it difficult to remember the attitudes and practices which prevailed when we turned our first sod at Hamilton in 1985. They have completely altered the land conservation landscape in Australia and are cause for optimism, despite the economic downturn faced by most farmers in the early 1990s.

We are only just realising the extent and implications of inappropriate land management—there is a long way to go before we have identified sustainable land uses for most agricultural land in Australia, even further before we have translated these into practical and profitable land management systems. There is a great deal more research to be done into the characteristics and capabilities of land systems in many areas, and into alternative enterprises, in particular to reduce reliance on external inputs of energy and pesticides. Of course, there is always room for more research, but there is a great deal which can be done in the meantime to make existing practices more sustainable. And it is important that ecological purists do not adopt the 'pull-the-ladder-up' position of expecting mainstream farmers using conventional practices to change overnight. Sustainability must incorporate economic and social components. It is more effective to get farmers themselves to question their existing systems and to begin to modify them than to tell them that they are 'stuffing up the country and have to change'.

If we can get all land users, every time they make a land management decision, to ask themselves: 'What are the implications of this action for the land now and in the future?' we will have taken a major step towards sustainable land use. One of the main benefits of whole farm planning, as applied on the Potter farms, is that it encourages farmers to ask this question. The degree to which the farms have moved along the path towards sustainable farming is for others to judge.

Index